Environmental Health Criteria 161

PHENOL

First draft prepared by Ms G.K. Montizaan,
National Institute of Public Health and
Environmental Hygiene, Bilthoven, Netherlands

Published under the joint sponsorship of
the United Nations Environment Programme,
the International Labour Organisation,
and the World Health Organization

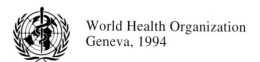

World Health Organization
Geneva, 1994

The **International Programme on Chemical Safety (IPCS)** is a joint venture of the United Nations Environment Programme, the International Labour Organisation, and the World Health Organization. The main objective of the IPCS is to carry out and disseminate evaluations of the effects of chemicals on human health and the quality of the environment. Supporting activities include the development of epidemiological, experimental laboratory, and risk-assessment methods that could produce internationally comparable results, and the development of manpower in the field of toxicology. Other activities carried out by the IPCS include the development of know-how for coping with chemical accidents, coordination of laboratory testing and epidemiological studies, and promotion of research on the mechanisms of the biological action of chemicals.

WHO Library Cataloguing in Publication Data

Phenol.

(Environmental health criteria ; 161)

1.Phenols - standards   2.Environmental exposure
I.Series

ISBN 92 4 157161 6   (NLM Classification: QD 341.P5)
ISSN 0250-863X

The World Health Organization welcomes requests for permission to reproduce or translate its publications, in part or in full. Applications and enquiries should be addressed to the Office of Publications, World Health Organization, Geneva, Switzerland, which will be glad to provide the latest information on any changes made to the text, plans for new editions, and reprints and translations already available.

CONTENTS

ENVIRONMENTAL HEALTH CRITERIA FOR
PHENOL

1. SUMMARY   11

    1.1 Identity, physical and chemical properties,
         analytical methods   11
    1.2 Sources of human and environmental exposure   11
    1.3 Environmental transport, distribution and
         transformation   11
    1.4 Environmental levels and human exposure   12
    1.5 Kinetics and metabolism   12
    1.6 Effects on laboratory mammals, and *in vitro* test
         systems   13
    1.7 Effects on humans   15
    1.8 Effects on organisms in the environment   15
    1.9 Summary of evaluation   16
         1.9.1   Human health   16
         1.9.2   Environment   17

2. IDENTITY, PHYSICAL AND CHEMICAL PROPERTIES,
    ANALYTICAL METHODS   18

    2.1 Identity   18
    2.2 Physical and chemical properties   18
    2.3 Conversion factors   20
    2.4 Analytical methods   20
         2.4.1   Sampling and pre-treatment   20
         2.4.2   Analysis   23

3. SOURCES OF HUMAN AND ENVIRONMENTAL
    EXPOSURE   25

    3.1 Natural sources   25
    3.2 Anthropogenic sources   25
         3.2.1   Production   25
         3.2.2   Industrial processes   25
         3.2.3   Non-industrial sources   27
    3.3 Endogenous sources   28
    3.4 Uses   28

4. ENVIRONMENTAL TRANSPORT, DISTRIBUTION AND
    TRANSFORMATION   30

    4.1 Transport and distribution between media   30
    4.2 Abiotic degradation   31
         4.2.1 Air   31
         4.2.2 Water   31
    4.3 Biodegradation   32

5. ENVIRONMENTAL LEVELS AND HUMAN EXPOSURE 34

    5.1 Environmental levels   34
         5.1.1 Air   34
         5.1.2 Water and sediment   34
    5.2 Occupational exposure   36
         5.2.1 Production   36
         5.2.2 Application of phenolic resins   36
         5.2.3 Other occupational situations   37
    5.3 General population exposure   38
         5.3.1 Indoor air   38
         5.3.2 Food and drinking-water   38

6. KINETICS AND METABOLISM IN LABORATORY
    ANIMALS AND HUMANS   40

    6.1 Absorption   40
         6.1.1 Animal uptake studies   40
                6.1.1.1 Pulmonary   40
                6.1.1.2 Dermal   40
                6.1.1.3 Intestinal   41
         6.1.2 Human uptake studies   41
                6.1.2.1 Pulmonary   41
                6.1.2.2 Dermal   41
    6.2 Distribution   42
    6.3 Metabolic transformation   43
         6.3.1 Metabolite identification   43
         6.3.2 Covalent binding to macromolecules   44
         6.3.3 Location   44
    6.4 Elimination and excretion   45
    6.5 Biological monitoring   46

7. EFFECTS ON LABORATORY MAMMALS, AND
    IN VITRO TEST SYSTEMS   48

    7.1 Single exposure   48

      7.1.1   LD$_{50}$ values   48
      7.1.2   Effects   48
7.2 Short-term exposure   50
      7.2.1   Oral exposure   50
      7.2.2   Dermal exposure   51
      7.2.3   Inhalation exposure   51
      7.2.4   Subcutaneous exposure   52
      7.2.5   Ear exposure   53
7.3 Skin and eye irritation; sensitization   53
7.4 Long-term exposure   54
7.5 Reproduction, embryotoxicity and teratogenicity   54
      7.5.1   Reproductive toxicity   54
      7.5.2   Embryotoxicity/teratogenicity   54
           7.5.2.1 *In vivo* studies   54
           7.5.2.2 *In vitro* studies   56
7.6 Mutagenicity and related end-points   57
      7.6.1   Mutagenicity studies   57
           7.6.1.1 Bacterial systems   57
           7.6.1.2 Non-mammalian eukaryotic systems   67
           7.6.1.3 Mammalian *in vitro* systems   67
           7.6.1.4 Mammalian *in vivo* systems:
                    somatic cells   68
           7.6.1.5 Mammalian *in vivo* systems:
                    germ cells   69
7.7 Carcinogenicity   70
      7.7.1   Oral exposure   70
      7.7.2   Dermal exposure   70
      7.7.3   Inhalation exposure   71
      7.7.4   Two-stage carcinogenicity studies   71
7.8 Special studies   71
      7.8.1   Neurotoxicity   71
      7.8.2   Myelotoxicity   72
      7.8.3   Immunotoxicology   73
      7.8.4   Biochemical effects   74

8.   EFFECTS ON HUMANS   75

8.1 General population exposure   75
      8.1.1   Controlled studies   75
      8.1.2   Case reports   75
           8.1.2.1 Dermal exposure   75
           8.1.2.2 Oral exposure   78
           8.1.2.3 Inhalation exposure   79
           8.1.2.4 Exposure by injection   79
8.2 Occupational exposure   79

8.3   Organoleptic data   81

9.   EFFECTS ON OTHER ORGANISMS IN THE
     LABORATORY AND FIELD   82

     9.1 Microorganisms   82
     9.2 Aquatic organisms   93
          9.2.1   Freshwater organisms   93
                  9.2.1.1   Short-term studies   93
                  9.2.1.2   Long-term studies   94
          9.2.2   Marine organisms   96
                  9.2.2.1   Short-term studies   96
                  9.2.2.2   Long-term studies   96
          9.2.3   Accumulation   96
          9.2.4   Metabolism   97
     9.3 Terrestrial organisms   98

10. EVALUATION OF HUMAN HEALTH RISKS AND
     EFFECTS ON THE ENVIRONMENT   99

     10.1 Evaluation of human health risks   99
          10.1.1   Exposure   99
          10.1.2   Toxicity   100
          10.1.3   Evaluation   101
     10.2 Evaluation of effects on the environment   102
          10.2.1   Environmental levels   102
          10.2.2   Toxicity   102
          10.2.3   Evaluation   103

11. FURTHER RESEARCH   104

12. PREVIOUS EVALUATIONS BY INTERNATIONAL
     BODIES   105

REFERENCES   107

RESUME   136

RESUMEN   144

WHO TASK GROUP ON ENVIRONMENTAL HEALTH CRITERIA FOR PHENOL

Members

Dr L.E. Hansen, dk-Teknik, Soeborg, Denmark

Dr R.J. Kavlock, Developmental Toxicology Division, Health Effects Research Laboratory, US Environmental Protection Agency, Research Triangle Park, North Carolina, USA

Dr C.J. Price, Neurotoxicology Program Development, Center for Life Sciences and Toxicology, Research Triangle Institute, Research Triangle Park, North Carolina, USA

Mr D. Renshaw, Department of Health, Elephant and Castle, London, United Kingdom

Dr A. Smith, Health and Safety Executive, Toxicology Unit, Bootle, Merseyside, United Kingdom (*Joint Rapporteur*)

Professor J.A. Sokal, Institute of Occupational Medicine and Environmental Health, Sosnowiec, Poland (*Chairman*)

Dr S.H.H. Swierenga, Health and Welfare Canada, Drugs Directorate, Ottawa, Ontario, Canada (*Joint Rapporteur*)

Dr T. Vermeire, National Institute of Public Health and Environmental Protection, Toxicology Advisory Centre, Bilthoven, The Netherlands

Secretariat

Professor F. Valić, IPCS Consultant, World Health Organization, Geneva, Switzerland, *also* Vice-Rector, University of Zagreb, Zagreb, Croatia (*Responsible Officer* and *Secretary*)

NOTE TO READERS OF THE CRITERIA MONOGRAPHS

Every effort has been made to present information in the criteria monographs as accurately as possible without unduly delaying their publication. In the interest of all users of the Environmental Health Criteria monographs, readers are kindly requested to communicate any errors that may have occurred to the Director of the International Programme on Chemical Safety, World Health Organization, Geneva, Switzerland, in order that they may be included in corrigenda.

* * *

A detailed data profile and a legal file can be obtained from the International Register of Potentially Toxic Chemicals, Case postale 356, 1219 Châtelaine, Geneva, Switzerland (Telephone No. 9799111).

* * *

This publication was made possible by grant number 5 U01 ES02617-14 from the National Institute of Environmental Health Sciences, National Institutes of Health, USA.

ENVIRONMENTAL HEALTH CRITERIA FOR PHENOL

A Task Group on Environmental Health Criteria for Phenol met at the British Industrial and Biological Research Association (BIBRA) Toxicology International, Carshalton, United Kingdom from 26 to 30 April 1993. Dr D. Anderson welcomed the participants on behalf of the host institution, and Professor F. Valić opened the Meeting on behalf of the three cooperating organizations of the IPCS (UNEP/ILO/WHO). The Task Group reviewed and revised the draft monograph and made an evaluation of the risks for human health and the environment from exposure to phenol.

The first draft of this monograph was prepared by Ms G.K. Montizaan, National Institute of Public Health and Environmental Hygiene, Bilthoven, the Netherlands.

Professor F. Valić was responsible for the overall scientific content of the monograph and for the organization of the meeting, and Dr P.G. Jenkins, IPCS, was responsible for the technical editing of the monograph.

The efforts of all who helped in the preparation and finalization of the monograph are gratefully acknowledged.

ABBREVIATIONS

DMBA	dimethylbenzathraline
EEC	European Economic Community
LOAEL	lowest-observed-adverse-effect level
MATC	maximum acceptable tolerance concentration
NOAEL	no-observed-adverse-effect level
NOEL	no-observed-effect level
NOLC	no-observed lethal concentration
PCE	polychromatic erythrocytes
TT	toxicity threshold

1. SUMMARY

1.1 Identity, physical and chemical properties, analytical methods

Phenol is a white crystalline solid which melts at 43 °C and liquefies upon contact with water. It has a characteristic acrid odour and a sharp burning taste. It is soluble in most organic solvents; its solubility in water is limited at room temperature; above 68 °C it is entirely water-soluble. Phenol is moderately volatile at room temperature. It is a weak acid, and in its ionized form very sensitive to electrophile substitution reactions and oxidation.

Phenol may be collected from environmental samples by absorption in NaOH solution or onto solid sorbents. Desorption is achieved by acidification, steam distillation and ether extraction (from solutions) or by thermal or liquid desorption (from solid sorbents). The most important analytical techniques are gas chromatography in combination with flame ionization/electron capture detection, and high-performance liquid chromatography in combination with ultraviolet detection. The lowest reported detection limit for air is 0.1 μg/m^3. Phenol can be measured in blood and urine; in urine samples a detection limit of 0.5 μg/litre has been reported.

1.2 Sources of human and environmental exposure

Phenol is a constituent of coal tar, and is formed during the natural decomposition of organic materials. The major part of phenol present in the environment, however, is of anthropogenic origin. Production and use of phenol and its products, especially phenolic resins and caprolactam, exhaust gases, residential wood burning and cigarette smoke are potential sources. Another potential source is the atmospheric degradation of benzene under the influence of light, whereas the presence of phenol in liquid manure may also contribute considerably to its atmospheric levels. Benzene and phenol derivatives may, by *in vivo* conversion, form a source of endogenous human phenol exposure.

The worldwide production of phenol appeared to be fairly constant throughout the 1980s, the USA being the most important producer. Its major use is as a feedstock for phenolic resins, bisphenol A and caprolactam. Some medical and pharmaceutical applications are also known.

1.3 Environmental transport, distribution and transformation

The main emissions of phenol occur to air. The major part of phenol in the atmosphere will be degraded by photochemical reactions to dihydroxybenzenes, nitrophenols and ring cleavage products, with an estimated half-life of 4-5 h. A minor part will disappear from the air by wet deposition (rain). Phenol is expected to be highly mobile in soil, but transport and reactivity may be affected by pH.

Phenol in water and soil may be degraded by abiotic reactions as well as microbial activity to a number of compounds, the most important being carbon dioxide and methane. The proportion of biodegradation to the overall degradation of phenol is determined by many factors, such as concentration, acclimation, temperature, and the presence of other compounds.

1.4 Environmental levels and human exposure

No data are available on atmospheric phenol levels. Background levels are expected to be less than 1 ng/m^3. Urban/suburban levels vary from 0.1 to 8 μg/m^3, while concentrations in source-dominated areas (industry) were reported to be up to two orders of magnitude higher. Phenol has been detected in rain, surface water and ground water, but data are very scarce. Elevated phenol levels have been reported in sediments and ground waters due to industrial pollution.

Occupational exposure to phenol may occur during the production of phenol and its products, during the application of phenolic resins (wood and iron/steel industry) and during a number of other industrial activities. The highest concentration (up to 88 mg/m^3) was reported for workers in the ex-USSR quenching coke with phenol-containing waste water. Most other reported concentrations did not exceed 19 mg/m^3.

For the general population, cigarette smoke and smoked food products are the most important sources of phenol exposure, apart from the exposure via air. Exposure by way of drinking-water and inadvertently contaminated food products should be low;

phenol has an objectionable smell and taste, which in most cases leads to non-acceptance by the consumer.

1.5 Kinetics and metabolism

Phenol is readily absorbed by all routes of exposure. After absorption, the substance is rapidly distributed to all tissues.

Absorbed phenol mainly conjugates with glucuronic acid and sulfuric acid and, to a lesser extent, hydroxylates into catechol and hydroquinone. Phosphate conjugation also occurs. The formation of reactive metabolites (4,4-biphenol and diphenoquinone) has been demonstrated in *in vitro* studies with activated human neutrophils and leucocytes.

The relative amounts of glucuronide and sulfate conjugates vary with dose and animal species. A shift from sulfation to glucuronidation was observed in rats after increasing the phenol dose.

The liver, the lung, and the gastrointestinal mucosa are the most important sites of phenol metabolism. The relative role played by these tissues depends on route of administration and dose.

In vivo and *in vitro* studies have demonstrated covalent binding of phenol to tissue and plasma proteins. Some phenol metabolites also bind to proteins.

Urinary excretion is the major route of phenol elimination in animals and humans. The rate of urinary excretion varies with dose, route of administration, and species. A minor part is excreted in the faeces and expired air.

1.6 Effects on laboratory mammals, and *in vitro* test systems

Phenol has moderate acute toxicity for mammals. Oral LD_{50} values in rodents range from 300 to 600 mg phenol/kg body weight. Dermal LD_{50} values for rats and rabbits range from 670 to 1400 mg/kg body weight, respectively, and the 8-h LC_{50} for rats by inhalation is more than 900 mg phenol/m^3. Clinical symptoms after acute exposure are neuromuscular hyperexcitability and severe convulsions, necrosis of skin and mucous membranes of the throat, and effects on lungs, nerve fibres, kidneys, liver, and the pupil response to light.

Solutions of phenol are corrosive to skin and eyes. Phenol vapours can irritate the respiratory tract. There is evidence that phenol is not a skin sensitizer.

The most important effects reported in short-term animal studies were neurotoxicity, liver and kidney damage, respiratory effects and growth retardation. Toxic effects in rat kidney have been reported to occur at oral dose levels of 40 mg/kg per day or more. Liver toxicity was evident in rats administered at least 100 mg/kg per day. In a limited 14-day study in rats, an oral no-observed-adverse-effect level (NOAEL) of 12 mg/kg per day was reported, based on kidney effects. In this experiment miosis (an iris response to light) was still inhibited at 4 mg/kg per day; however, the health significance of this finding is not clear. Some biological changes were reported to occur in the intestinal mucosa and kidneys of mice at dose levels below 1 mg/kg per day, a finding of uncertain toxicological significance.

There are no adequate studies on the reproductive toxicity of phenol. Phenol has been identified as a developmental toxicant in studies with rats and mice. In two multiple dose rat studies, NOAEL values of 40 mg/kg per day (the lowest-observed-adverse-effect level (LOAEL) was 53 mg/kg per day) and 60 mg/kg per day (the LOAEL was 120 mg/kg per day) have been reported. In the mouse, the NOAEL was 140 mg/kg per day (the LOAEL was 280 mg/kg per day).

The majority of bacterial mutagenicity tests have given negative results. Mutations, chromosomal damage and DNA effects have been observed in mammalian cells *in vitro*. Phenol has no effect on intercellular communication (measured as metabolic cooperation) in cultured mammalian cells. Induction of micro-nuclei in bone marrow cells of mice has been observed in some studies. No micronuclei were observed in mice studies at lower doses.

Two carcinogenicity studies have been conducted with male and female rats and mice receiving phenol in their drinking-water. Malignancies (e.g., C-cell thyroid carcinoma, leukaemia) were only seen in low-dose male rats. No adequate dermal or inhalation carcinogenicity studies have been conducted. Two-stage carcinogenicity studies have shown that phenol, applied repeatedly to mouse skin, has promoting activity.

1.7 Effects on humans

A wide range of adverse effects has been reported following well-documented human exposure to phenol by the dermal, oral or intravenous routes. Gastrointestinal irritation has been reported following ingestion. Local effects following dermal exposure range from painless blanching or erythema to corrosion and deep necrosis. Systemic effects include cardiac dysrhythmias, metabolic acidosis, hyperventilation, respiratory distress, acute renal failure, renal damage, dark urine, methaemoglobinaemia, neurological effects (including convulsions), cardiovascular shock, coma and death. The lowest reported dose resulting in a human death was 4.8 g by ingestion; death occurred within 10 min.

The potential for poisoning through inhalation of phenol vapours has long been recognized, but no cases of death following this route of exposure have been reported. Symptoms associated with inhalation of phenol included anorexia, weight loss, headache, vertigo, salivation and dark urine.

Phenol is not a sensitizing agent.

The human odour threshold for phenol has been reported to range from 0.021 to 20 mg/m^3 in air. The odour threshold for phenol in water has been reported to be 7.9 mg/litre, and the taste threshold 0.3 mg/litre in water.

Adequate human data on the carcinogenicity of phenol are not available.

1.8 Effects on organisms in the environment

In studies on single bacteria species, the EC_{50} values found for growth inhibition varied from 244 to 1600 mg phenol/litre. A toxicity threshold of 64 mg phenol/litre was found. Values for protozoa and fungi were of the same order of magnitude as for bacteria; for algae, they were somewhat lower.

Phenol is toxic to higher freshwater organisms. The lowest LC_{50} or EC_{50} values for crustaceans and fish lie between 3 and 7 mg phenol/litre. The data on the acute toxicity to marine organisms are comparable with those for freshwater organisms. In long-term studies on crustacea and fish species, a remarkable difference in sensitivity has been observed; the LC_1 values from embryo-larval tests on *Salmo* and *Carassius* proved to be much lower (0.2 and 2 μg phenol/litre, respectively) than the corresponding values for other fish species (NOLC 2.2-6.1 mg/litre) and amphibia, or from reproduction tests on crustacea (NOLC 10 mg phenol/litre). Data from long-term tests on marine organisms are not available.

The bioconcentration factors of phenol in various types of aquatic organisms are in general very low (< 1-10), although some higher values (up to 2200) have also been reported. Phenol, therefore, is not expected to bioaccumulate significantly.

The available data concerning the fate and effects of phenol in terrestrial organisms are very scarce. A 120-h EC_{50} for millet was found to be 120-170 mg/litre, and in a contact test the LC_{50} for earthworm species was 2.4-10.6 μg/cm².

1.9 Summary of evaluation

1.9.1 Human health

The general population is primarily exposed to phenol by inhalation. Repeated oral exposure may arise from consumption of smoked food or drinking-water.

Data are inadequate to determine the degree of exposure of the general population, but an upper-limit estimate of the daily intake can be made. On the basis of "the worst case scenario", an estimate can be made assuming that an individual will be maximally exposed to phenol through continuous inhalation of heavily contaminated air with frequent consumption of smoked food items and of drinking-water containing phenol up to the taste threshold. The estimated maximal total daily intake of phenol for such a 70-kg individual is calculated to be 0.1 mg/kg body weight per day.

The lowest NOAEL values identified in animal experiments are for kidney and developmental effects, and in rats lie within the range of 12-40 mg/kg body weight per day. Using an uncertainty factor of 200, a range of 60-200 μg/kg body weight per day is recommended as the upper limit of the total daily intake (TDI). Considering the upper-limit estimate of human daily intake of 100 μg/kg body weight per day, it is concluded that the average general population exposure to phenol from all sources is below this range.

A reason for concern is some evidence that phenol may be genotoxic and the fact that there is insufficient data to discount with certainty the possible carcinogenicity of the compound. The evaluation must be kept under periodic review.

1.9.2 Environment

Phenol is not expected to bioaccumulate significantly. Phenol is toxic to aquatic organisms; an environmental concern level of 0.02 μg/litre can be determined applying the modified US EPA method. Adequate data on plants and terrestrial organisms are lacking.

Intercompartmental transport of phenol mainly occurs by wet deposition and by leaching through soil. Generally, the compound is not likely to persist in the environment. The scarce exposure data do not allow the evaluation of the risk from phenol to either aquatic or terrestrial ecosystems. However, in view of the derived environmental concern level for water, it is reasonable to assume that aquatic organisms may be at risk in any surface or sea water contaminated with phenol.

2. IDENTITY, PHYSICAL AND CHEMICAL PROPERTIES, ANALYTICAL METHODS

2.1 Identity

Chemical formula: C_6H_6O

Chemical structure:

OH

Relative molecular
mass: 94.11

Common name: phenol

Common synonyms: acidum carbolicum, acidum phenolicum, acidum phenylicum, benzaphenol, benzene phenol, benzenol carbolic acid, hydroxybenzene (IUPAC), oxybenzene, monohydroxybenzene, monophenol, phenic acid, phenol alcohol, phenyl hydrate, phenyl hydroxide, phenylic acid

Common trade names: carbololie (NLD), fenololie (NLD), kristalliertes Kreosot (GER), Steinkohlenkreosot (GER), Steinkohlenteerkreosot (GER), venzénol (FRA), ENT 1814.

CAS registry
number: 108-95-2

CAS chemical name: phenol

2.2 Physical and chemical properties

Some physical and chemical properties of phenol are given in Table 1.

Table 1. Some physical and chemical properties of phenol[a]

Boiling point (101.3 Pa)	181.75 °C
Melting point	43 °C
	40.9 °C (ultrapure material)
Relative density (20 °/4 °)[b]	1.071
Relative vapour density (air = 1)	3.24
Vapour pressure (20 °C)	0.357 mmHg
(50 °C)	2.48 mmHg
(100 °C)	41.3 mmHg
Saturation concentration in air (20 °C)	0.77 g/m^3
Solubility in water (16 °C)	67 g/litre[c]
Log *n*-octanol/water partition coefficient (log P_{ow})	1.46[d]
Dissociation constant in water at 20 °C (K_a)	1.28 x 10^{-10}
Flash-point (closed cup)	80 °C
(open cup)	79 °C
	85 °C[e]
Flammability limits	1.3-9.5%

[a]   From: Kirk-Othmer (1980); RIVM (1986)
[b]   Weast (1987)
[c]   Above 68.4 °C phenol is entirely soluble in water
[d]   The P_{ow} of phenol is very much dependent on pH; pH at log P_{ow} = 1.46 was not given
[e]   Budavari et al. (1989)

Phenol has a melting point of 43 °C and forms white to colourless crystals (Budavari et al., 1989). It has also been described as a colourless to pink solid or thick liquid (NIOSH, 1985a). Phenol has a characteristic acrid smell and a sharp burning taste. Odour and taste threshold values are reported in section 8.3. In the molten state, it is a clear, colourless liquid with a low viscosity. A solution with approximately 10% water is called phenolum liquefactum, as this mixture is liquid at room temperature. Phenol is soluble in most organic solvents (aromatic hydrocarbons, alcohols, ketones, ethers, acids, halogenated hydrocarbons). The solubility is limited in aliphatic solvents.

The chemical properties of phenol are affected by the resonance stabilization possibilities of phenol and, in particular, of the phenolate ion. Because of this, phenol reacts as a mild acid. In the presence of electrophilic groups (meta-indicators), the acidic properties are emphasized.

Phenol is sensitive to oxidizing agents. Splitting of the hydrogen atom from the phenolic hydroxyl group is followed by resonance stabilization of the resulting phenyloxy radical. The radical formed can easily be further oxidized. Depending on the oxidizing agent applied and the reaction conditions, various products, such as dihydroxy- and trihydroxybenzenes and quinones are formed. These properties make phenol suitable as an antioxidant, functioning as a radical trapping agent. Phenol undergoes numerous electrophilic substitution reactions, such as halogenation and sulfonation. It also reacts with carbonyl compounds in both acidic and alkaline media. In the presence of formaldehyde, phenol is readily hydroxymethylated with subsequent condensation to resins.

2.3 Conversion factors

1 mg/m^3 = 0.26 ppm
1 ppm   = 3.84 mg/m^3

2.4 Analytical methods

Analytical methods for phenol are shown in Table 2.

2.4.1 Sampling and pre-treatment

Phenol in air samples may be collected by absorption in NaOH solution contained in wash bottles or on filters impregnated with NaOH solution. Phenol in air, water and solid waste samples may be collected (directly or after extraction) in tubes containing solid sorbent (Tenax, silica gel or, less commonly, carbon) (IARC, 1989). For large air volumes, the NaOH method is usually preferred, whereas for smaller quantities (personal air sampling, for instance) solid sorbent tubes have been reported to be more practical (RIVM, 1986).

Table 2. Methods for the detection of phenol in air

Sampling	Volume of air (litre)	Pre-treatment before analysis	Analysis	Detection limit ($\mu g/m^3$)	Reference
Absorbance in NaOH solution in a wash bottle; 1 litre/min	100	acidification	GC and FID	10 μg per sample	NIOSH (1984)
Absorbance in NaOH solution in an impinger; 20 litre/min	25 000	acidification and steam distillation	GC and FID	4	Katz (1977)
Glass fibre filter impregnated with NaOH and glycerol; 120 litre/min	600	acidification and extraction with ether	GC and FID	13	Kifune (1979)
Absorbance in NaOH solution in an impinger; 28 litre/min	1000	acidification and steam distillation	colorimetry with 4-amino antipyrine	2	Katz (1977)
Absorbance in NaOH solution in a wash bottle; 1 litre/min	-	none	colorimetry with 4-amino antipyrine	700	Hensehler (1975)
Absorbance in NaOH solution in a sinter wash bottle; 1-2 litre/min	150	conversion in an azophenol derivate	HPLC with UV detector (254 nm)	0.2	Kuwata et al. (1980)
Absorbance in Na_2CO_3 solution; 1 litre/min	30	calibration at pH = 10 and pH = 7	UV at 235 nm at 2 pH values	160	Zavorovskaya & Nekhorosheva (1981)

Table 2 (contd).

Sampling	Volume of air (litre)	Pre-treatment before analysis	Analysis	Detection limit ($\mu g/m^3$)	Reference
Absorbance in NaOH solution in an impinger; 28 litre/min	300	calibration at pH = 12 and at pH = 6	UV at 241 and 295 nm	20	Bergshoeff (1960)
Tenax with and without KOH; 0.25 litre/min	5	thermal desorption; 250 °C	GC and FID	0.1	Hoshika & Muto (1979)
Tenax; 0.1 litre/min	4	thermal desorption; 260 °C	GC and FID	1	Russell (1975)
Tenax; 0.5-1 litre/min	70	thermal desorption; 250 °C	GC and MS	0.3	Hagemann et al. (1978)
Tenax; 0.75 litre/min	90	thermal desorption; 300 °C	IR with 20 m gas cuvette	300	Podolak et al. (1981)
Silica gel	25	liquid desorption with chloroform	GC and FID	3	Dimitriev & Mishchikhin (1983)
Silica gel; 0.05-2 litre/min	10	liquid desorption with chloroform	HPLC with UV detector (275 nm)	50	Oomems & Schuurhuis (1983)
Drägen tube gas detection	0.5	none	reading of colorisation	19 000	Leichnitz (1982)

22

Release of phenol from aqueous solutions (including NaOH sorbent, and also urine) is achieved by acidification, steam distillation and/or ether extraction. After adsorption onto Tenax, thermal desorption at 250 °C is usually preferred (the whole sample may be inserted directly into a gas chromatograph), whereas, in the case of silica gel, liquid desorption with chloroform is generally applied. There is a small possibility of chemical conversion during heating, whereas Tenax may react with ozone to form small quantities of phenol. Only one analysis per sample is possible in the case of thermal desorption. The use of liquid desorption allows more analyses per sample, but because of the unavoidable dilution the detection limit is higher (RIVM, 1986).

2.4.2 Analysis

The most important analytical techniques for the detection of phenol are gas chromatography (GC) in combination with flame ionization detection (FID), and high-performance liquid chromatography (HPLC) in combination with ultraviolet (UV) detection. The identification of phenol by GC/FID has been improved by reaction of phenols with bromide or pentafluoro-benzyl bromide, and the use of electron capture detection (Hoshika & Muto, 1979; US EPA, 1986a). The identification of phenol using HPLC can be improved by reaction with, for example, p-nitrobenzene diazonium tetrafluoroborate to form azo derivatives (Kuwata et al., 1980).

Detection limits of the above techniques for air samples are given in Table 2. For the GC/FID detection of phenol in water, using electron capture detection following derivatization with pentafluorobenzyl bromide, a detection limit of approximately 0.2 μg/litre has been reported (US EPA, 1986a).

GC in combination with mass spectrometry (MS) is more sensitive than with FID, but is more expensive. This technique, using either packed or capillary columns, was reported to have practical quantitative limits of approximately 1 mg phenol/kg wet weight for soil/sediment samples, 1-200 mg phenol/kg for wastes, and 10 μg/litre for groundwater samples (US EPA, 1986b,c).

Another reported analytical technique is colorimetry after reaction of phenol with 4-amino antipyrine, in the presence of potassium ferricyanide, to form an antipyrine dye. The detection limit of this technique for water samples, after steam distillation, was reported to be 1 μg/litre (American Public Health Association, 1985). For air samples of 1 m^3, the detection limit was reported to be 2 μg/m^3 (see Table 2). The interference by para-substituted phenols and chlorophenols is low (RIVM, 1986).

Infrared (IR) detection of phenol is a rather insensitive method, and is highly susceptible to interference by other compounds such as water vapour. However, it is a rapid and specific method which allows directly readable continuous measurement. It is considered to be attractive only at air concentrations of more than 1000 μg/m^3, for example in leakage tests and industrial warning systems (RIVM, 1986). Also directly readable is the Dräger gas detection tube (see Table 2); however, the detection limit is very high (> 19 000 μg/m^3).

For the GC/FID detection of total phenol in urine samples, after acidification and ether extraction, a detection limit of 0.5 μg/litre was estimated (NIOSH, 1985b). Colorimetric methods for the determination of free phenol in both urine and blood are available. In one method, phenol reacts with *p*-nitroaniline following deproteinization and extraction with diethyl ether. Other phenols will interfere (Müting et al., 1970). In another method, phenol reacts with ammonia and *N*-chlorosuccinimide in alkaline media with sodium nitroprusside as a catalyst. This method was found to be applicable in the range of 3-24 mg/litre, using spiked samples of urine (Amlathe et al., 1987). The concentration of total phenol in urine and plasma can be determined by GC/MS following hydrolysis of glucuronide and sulfate conjugates with sulfuric acid and derivatization with propanoic anhydride. The detection limit is reported to be 10 μg/litre (Pierce & Nerland, 1988).

3. SOURCES OF HUMAN AND ENVIRONMENTAL EXPOSURE

3.1 Natural sources

Phenol is a constituent of coal tar, and is formed during the natural decomposition of organic materials. Increased environmental levels may result from forest fires (Hubble et al., 1981).

Phenol has been detected among the volatile components from liquid manure at concentrations of 7-55 μg/kg dry weight (Spoelstra, 1978). In the Netherlands, for example, the contribution from this source to the overall phenol emission into air in 1983 has been calculated to be 15%, assuming complete volatilization of phenol and an average phenol concentration in manure of 30 μg/kg dry weight (RIVM, 1986).

3.2 Anthropogenic sources

3.2.1 Production

The most commonly used production method for phenol, on a worldwide scale, is from cumene (isopropylbenzene). In the USA, for example, more than 98% of phenol is produced by this method (IARC, 1989). Phenol is also produced from chlorobenzene and toluene. A small but steady supply of phenol is recovered as a by-product of metallurgical coke manufacture (IARC, 1989). The emission factor of phenol into air during production by the cumene process has been reported to be 0.16 g phenol emitted per kg phenol produced (UBA, 1981).

In Table 3, information is presented concerning the production of phenol in various countries in 1981. This information was derived from the open literature (Chemfacts, 1978-1981; United Nations, 1980; SRI, 1982; CID-TNO, 1984; IARC, 1989) and, where necessary, was extrapolated to 1981. There have been no major production changes according to data available up to 1986 (IARC, 1989).

3.2.2 Industrial processes

Phenol is the basic feedstock from which a number of commercially important materials are made, including phenolic

Table 3. Production of phenol in 1981 and 1986 (kilotonne/year)

Country	Production in 1981[a]	Production in 1986[b]
Brazil	50	
Bulgaria	35	
Czechoslovakia	44	46
Finland	32	
France	150	
Germany, Federal Republic of	247	
Italy	223	
India	14	
Japan	215	260
Mexico	20	
The Netherlands	166	
Poland	66	
Romania	66	
Spain	55	70
United Kingdom[c]	110	53
USA	1350	1413
USSR	497	515
Other countries	34	
European Community (total)	920	
Total	3374	

[a] Chemfacts (1978-1981); United Nations (1980); SRI (1982); CID-TNO (1984); IARC (1989)
[b] From IARC (1989)
[c] Phenol is no longer produced in the UK

resins, bisphenol A (2,2-bis-1-hydroxyphenylpropane), capro-lactam, alkyl phenols, as well as chlorophenols such as pentachlorophenol (IARC, 1989).

The most important phenol emissions result from the use of phenolic resins. Phenolic resins are used as a binding material in, for example, insulation materials, chipboard and triplex, paints and casting sand foundries. Their contents vary from 2-3% for insulation material to > 50% for moulds (Bollig & Decker, 1980). Emissions are approximately proportional to the concentration of free phenol, which is present as a monomer in these materials (1-5%) (Bollig & Decker, 1980). In addition, phenol may be released as a result of thermal decomposition of the resins.

In foundries, phenol emissions develop both during the production of moulds and kernels and during founding (TNO, 1978). The content of free phenol may rise by up to 12% (Ryser & Ulmer, 1980). Emission factors reported by RIVM (1986) were 0.35 g phenol emitted per kg used casting sand, 2-5 g phenol emitted per kg resin in the production of casting sand, and 10 g phenol emitted per kg resin during the production of moulds by the "hot-box" procedure.

Other industrial activities in which phenol may be emitted to the air, as well as some of their reported emission factors, are listed below:

- production of phenol resins (0-0.5 g phenol emitted per kg resin produced) (RIVM, 1986)
- production of phenols and phenol derivatives
- production of caprolactam (0.02-0.05 g phenol emitted per kg cyclohexanone (an intermediate) produced (RIVM, 1986)
- production of cokes
- production of insulation materials
- process emissions

Emissions to water may also result from processing.

3.2.3 Non-industrial sources

Phenol has been detected in the exhaust gases of private cars at concentrations of 0.3 ppm (approximately 1.2 mg/m^3) to 1.4-2.0 ppm (5.4-7.7 mg/m^3) (Kuwata et al., 1980; Verschueren, 1983). It has also been identified in cigarette smoke, in quantities that are comparable to an average emission of 0.4 mg/cigarette (Groenen, 1978). Emission gases from all material incinerators and home fires, especially wood-burning, may contain substantial quantities of phenol (Den Boeft et al., 1984).

Another potential source of phenol is the atmospheric degradation of benzene under the influence of light (Hoshino & Akimoto, 1978).

Phenols have been detected in smoked foods (section 5.3.2).

3.3 Endogenous sources

An important additional source of human phenol exposure may be the *in vivo* formation from various xenobiotics, e.g., benzene (Pekari et al., 1992).

3.4 Uses

The largest single use of phenol is the production of phenolic resins. Next is its use in the production of caprolactam, an intermediate in the production of nylon 6, and 2,2-bis-1-hydroxyphenylpropane (bisphenol A), which is mainly used in the production of phenolic resins (Kirk-Othmer, 1980).

The various applications of phenol as a percentage of total 1981 consumption, in the USA and western Europe, are summarized in Table 4 (Kirk-Othmer, 1980). The data presented are in close agreement with the 1986 USA data reported by IARC (1989).

Table 4. Use of phenol in 1981 (% total consumption)[a]

Production of	USA	West Europe
Phenolic resins	48	36
Bisphenol A(2,2-bis-1-hydroxyphenylpropane)	18	17
Caprolactam	15	28
Other products	19	19
Total	100	100

[a] From: Kirk-Othmer (1980)

Phenol was widely used in the 19th century for wound treatment and as an antiseptic and local anaesthetic. The medical uses of phenol today include incorporation into disinfectants, antiseptics, lotions, salves and ointments (IARC, 1989). Another medical application of phenol is its use as a neurolytic agent, applied in order to relieve spasm and chronic pain (Wood, 1978).

In addition to the applications mentioned in section 3.2.2, phenol is used in the manufacture of paint and varnish removers, lacquers, paints, rubber, ink, illuminating gases, tanning dyes, perfumes, soaps and toys (IARC, 1989).

4. ENVIRONMENTAL TRANSPORT, DISTRIBUTION AND TRANSFORMATION

4.1 Transport and distribution between media

No data have been found concerning wet and dry deposition of phenol. Since phenol in air is present almost exclusively in the gas phase, dry deposition (by particle deposition) is expected to be negligible. Wet deposition may contribute to the disappearance of phenol from the atmosphere: when phenol was measured during seven episodes of rain in Portland, Oregon, USA, relatively high concentrations were found in the rain water (Leuenberger et al., 1985).

Based on its relatively high solubility in water and the relatively low vapour pressure at room temperature, phenol is expected to end up largely in the water phase upon distribution between air and water. Consequently, transport from air to soil and water is likely (RIVM, 1986). Volatilization from dry near-surface soil should be relatively rapid (Howard, 1989).

Theoretical deposition rates for phenol were estimated assuming a behaviour similar to SO_2, and comparing with the rate of reaction of phenol with hydroxyl radicals (see below). Based on this comparison, it was concluded that most phenol in the atmosphere is degraded chemically, rather than transported (RIVM, 1986).

Partition coefficient (K_{oc}) values of phenol for two silt loams were reported to be 39 and 91 dm^3/kg. Based on these K_{oc} values, phenol would be expected to be highly mobile in soil, and therefore may leach to ground water (Howard, 1989). This was confirmed by Scott et al. (1982) who found low adsorption of phenol to two sterile silt loams (pH 5.4, organic matter content 1.1 and 3.6, respectively), as shown by Freundlich K values of 0.57 and 1.19, respectively. Based on the pK_a (log (1.28 x 10 $^{-10}$)), phenol exists in a partially dissociated state in water and moist soils and, therefore, its transport and reactivity may be affected by pH (Howard, 1989). Upon measurement of the sorption and desorption of phenol from water to surface sediment (pH 6.21-6.35; organic matter content of fine fraction (< 2 μM) was 10.2%), phenol appeared to bind strongly to the soil. The estimated K_{oc} was 2900 dm^3/kg (Isaacson & Frink, 1984). However, no correction was made for any degradation occurring during the experiments. The adsorption of phenol onto soil or microbial biomass may be decreased by the presence of phenol derivatives (Boyd, 1982; Selvakumar & Hsieh, 1988). Phenol has been detected in ground water as a result of leaching (see section 5.1.2).

4.2 Abiotic degradation

4.2.1 Air

Phenol may react in air with hydroxyl and NO_3 radicals, and undergo other photochemical reactions to form dihydroxy-benzenes, nitrophenols, and ring cleavage products (Atkinson et al., 1979; Bruce et al., 1987). The half-life for phenol in air was found to be 4-5 h under photochemically reactive conditions in a smog chamber (Spicer et al., 1985); this is in good agreement with the estimated half-life of phenol in air of 5 h based on its estimated reaction rate with hydroxyl radicals (RIVM, 1986). Howard (1989) reported an estimated half-life of 15 h for the reaction of phenol with hydroxyl radicals in air. The reaction of phenol with nitrate radicals during the night may be a significant removal process; a half-life of 15 min has been estimated at an atmospheric concentration of 2×10^8 nitrate radicals per cm^3 (Howard, 1989).

Phenol absorbs light in the region of 290-330 nm and therefore could photolyse (Howard, 1989).

4.2.2 Water

Phenols generally react in sunlit natural water via reaction with photochemically produced hydroxyl and peroxy radicals; typical half-lives were reported to be 100 and 19.2 h, respectively (Howard, 1989).

Phenol was found to be oxidized to carbon dioxide in water under experimental conditions (temperature approximately 50 °C), in the presence of oxygen and sunlight, at a rate of 11% per 24 h (Knoevenagel & Himmelreich, 1976). It was reported to react with nitrate ions in dilute aqueous solutions to form dihydroxybenzenes, nitrophenols, nitrosophenol and nitroquinone, presumably by a radical mechanism involving hydroxyl and phenoxyl radicals (Niessen et al., 1988). Phenol has been found to react with nitrous acid in waste water to form cyanide (Adachi et al., 1987), and to form chlorophenols in chlorinated drinking-water (Jarvis et al., 1985) and *p*-benzoquinone in the presence of chlorine dioxide (Wajon et al., 1982).

4.3 Biodegradation

Bacteria play a major role in the degradation of phenol in soil, sediment and water. The number of bacteria capable of utilizing phenol is usually a small percentage of the total population present in, for example, a soil sample (Hickman & Novak, 1989). However, repeated phenol exposure may result in acclimation (the promotion of strains capable of utilizing phenol as food) (Young & Rivera, 1985; Colvin & Rozich, 1986; Shimp & Pfaender, 1987; Wiggins & Alexander, 1988; Tibbles & Baecker, 1989).

Phenol may be converted by bacteria under aerobic conditions to carbon dioxide (Southworth et al., 1985; Ursin, 1985; Aelion et al., 1987; Dobbins et al., 1987; Aquino et al., 1988), and under anaerobic conditions to carbon dioxide (Bak & Widdell, 1986; Tschech & Fuchs, 1987) or methane (Healy & Young, 1979; Ehrlich et al., 1982; Young & Rivera, 1985; Fedorak & Hrudey, 1986; Fedorak et al., 1986). Benzoate, catechol, *cis-cis*-muconate, β-ketoadipate, succinate and acetate have all been identified as intermediates in the biodegradation of phenol (Paris et al., 1982; Krug et al., 1985; Fedorak et al., 1986; Knoll & Winter, 1987). Some of the carbon derived from the degradation of phenol may be incorporated into the bacterial biomass (Chesney et al., 1985).

Phenol may be degraded in its free form as well as after adsorption onto soil or sediment, although the presence of sorbent reduces the rate of biodegradation (Shimp & Young, 1987; Knezovich et al., 1988).

When phenol is the only carbon source, it can be degraded in a biofilm reactor with first-order kinetics at concentrations below about 20 μg/litre at 10 °C. The first-order rate constants are 3 to 30 times higher than those of easily degraded organic compounds at 100- to 1000-fold higher concentrations (Arvin et al., 1991). Reported phenol degradation rates suggest rapid aerobic degradation in sewage (typically > 90% with an 8-h retention time), soil (typically complete biodegradation in 2-5 days), fresh water (typically complete biodegradation in < 1 day), and sea water (typically 50% in 9 days) (Howard, 1989). Anaerobic biodegradation is slower (Baker & Mayfield, 1980).

The contribution of bacteria to the overall rate of degradation may be affected by a number of factors such as phenol concentration (Baker & Mayfield, 1980; Ursin, 1985; Hwang et al., 1989), temperature (Baker & Mayfield, 1980; Bak & Widdell, 1986; Hwang et al., 1986; Thornton-Manning et al, 1987; Gurujeyalashmi & Oriel, 1989), sunlight (Hwang et al., 1986), soil depth (Dobbins et al., 1987; Federle, 1988), the presence of other nutrients required for bacterial growth (Rubin & Alexander, 1983; Fedorak & Hrudey, 1986; Rozich & Colvin, 1986; Thorton-Manning et al., 1987), the presence of other pollutants (Southworth et al., 1985; Hoffmann & Vogt, 1988; Wang et al., 1988; Namkoong et al., 1989) and bacterial abundance (Tranvik et al., 1991).

5. ENVIRONMENTAL LEVELS AND HUMAN EXPOSURE

5.1 Environmental levels

5.1.1 Air

No data are available for background levels of phenol in air, away from emission sources. They are expected to be low (< 1 ng phenol/m^3) (RIVM, 1986).

Higher levels of phenol in air may be expected for urban areas, mainly due to traffic emissions. Urban phenol concentrations have been reported for Osaka, Japan (1-4 μg phenol/m^3; Kuwata et al., 1980), Nagoya, Japan (0.2-8 μg phenol/m^3 with an average of 1.7 μg phenol/m^3; Hoshika & Muto, 1979, 1980), Paris, France (0.7-8 μg phenol/m^3; Hagemann et al., 1978), and Portland, USA (0.22 to 0.42 μg phenol/m^3; Leuenberger et al., 1985). Despite differences in analytical techniques, the first three series of measurements showed good agreement. The Portland results were lower, but came from air samples taken during rain periods; phenol was also detected in rain water (see section 5.1.2).

Ambient air levels of phenol have been extensively monitored in the highly industrialized and urbanized Upper Silesia region of Poland (Sanitary Epidemiological Station, Katowice, 1991). Levels during 1990 ranged from 3.8 to 26.6 μg/m^3, the highest values being in the areas of greatest industrial concentration.

Hoshika & Muto (1979, 1980) reported a phenol level of approximately 190 μg/m^3 "near" a phenolic resin factory (no details). Kuwata et al. (1980) found phenol levels of 0.8-3.5 mg/m^3 in foundry emissions (no details).

Based on limited data, median ambient atmospheric levels of phenol (based on estimated 24-h averages) were estimated by Brodzinsky & Singh (1982) to be 0.12 μg/m^3 for urban/suburban areas (7 samples) in the USA (which is lower than reported above for several cities), and 104 μg/m^3 (2-170 μg/m^3) for source-dominated areas (83 samples) in the USA.

5.1.2 Water and sediment

Levels of phenol dissolved in rain water from Portland, USA, were found to range from 0.08 to 1.2 μg/litre and averaged above 0.28 μg/litre; gas phase concentrations ranged from 220-410 ng/m³ (Leuenberger et al., 1985).

Concentrations reported for surface water in the Netherlands were 2.5-6.5 μg/litre for two major rivers, 0.3-7 μg/litre for lakes, and 1.5 μg/litre for coastal waters (the given concentrations include other phenolic substances) (RIVM, 1986). Industrial rivers in the USA were reported to contain 0-5 μg/litre, but 3-24 μg/litre was reported for Lake Huron. Phenol was also detected in 2/100 raw water supplies in 1977 in the US EPA National Organics Monitoring Survey (Howard, 1989).

Drinking-water levels of phenol in the USA have been reported to be around 1 μg/litre or otherwise below the detection limit (summarized by Howard, 1989). Phenol was detected (no quantitative data) in drinking-water in the USA from 5 out of 14 drinking-water plants surveyed and in Great Britain in 2 out of 4 sites (Fielding et al., 1981). Higher groundwater levels have been reported following industrial activity (e.g. 6.5-10 000 μg/litre in two aquifers 15 months after a coal gasification project; summarized by Howard, 1989). Phenol was detected at a maximum concentration of 1130 mg/litre in nine wells in Wisconsin after a spill, and was detectable for at least 1.5 years after the spill (Delfino & Dube, 1976).

Phenol was not detected in water samples from three areas in Japan analysed for an environmental survey; however, levels of 0.03-0.04 mg/litre were detected in 3 out of 9 bottom sediment samples from the same regions (Fujii, 1978).

Other sediment concentrations reported were 13 000 μg/kg in samples from Lake Huron, "not detected" in an unspecified industrial river in the USA, < 1000 μg/kg (dry weight) as the median concentration in 9% of sediment samples from 318 data points in the USA, and 10 μg/kg (dry weight) in samples collected 6 km from a wastewater treatment discharge zone in California (summarized by Howard, 1989).

Phenol was detected in 63 out of 165 sediments from sampling areas in the Puget Sound region (Tetra Tech. Inc., 1986). Half the samples had a concentration of phenol below 40 μg/kg sediment (dry weight); the maximum level was 1700 μg/kg.

Levels of phenol with means of 0.01-5.7 mg/litre (maximum up to 53 mg/litre) have been reported in effluents from various industrial sources (summarized by Howard, 1989). Highest levels were associated with the iron and steel industry. Limited quantitative data from the VIEW Database (ATSDR, 1989) for ground water at hazardous waste sites indicated maximum levels of 2.48 to 85 000 μg/litre (average 33 800 μg/litre, 6 data points).

No data have been found indicating the presence of phenol in soil. Phenol is not likely to persist in soil because of rapid biodegradation (section 4.3) or transport to ground water or air (section 4.1).

5.2 Occupational exposure

Occupational exposure to phenol may occur during the production of phenol and phenol derivatives, during the production of phenolic resins and other products derived from phenol, during processing of the latter materials, and during a number of other activities.

5.2.1 Production

Personal air samples of workers involved in the production of phenol by the cumene process in the ex-USSR contained on average 5.8 mg phenol/m^3. For workers occupied in the production of phenol from chlorobenzene, the mean exposure level was 1.2 mg phenol/m^3 (Mogilnicka & Piotrowski, 1974). Values reported in the same publication for workers in two phenol resin-producing industries were 0.6-3 mg phenol/m^3.

5.2.2 Application of phenolic resins

Occupational exposure during the processing of phenolic resins appears to be partly determined by the content of free phenol in the applied resin (Bollig & Decker, 1980; Ryser & Ulmer, 1980).

In the wood industry, indoor phenol concentrations of 0.3 mg per m^3 (Winkler, 1981) and 1.5 mg/m^3 (range 0.8-2.6) (Gilli et al., 1980) have been reported. Concentrations in the breathing zone of wood workers were 1.3-2.6 mg phenol/m^3 (Gspan et al., 1984).

In another study, concentrations of < 0.04 to 1.9 mg phenol/m³ were found at plywood plants (Kauppinen et al., 1986).

In iron and steel foundries, average hourly phenol concentrations of 0.4-4.5 mg/m³ were reported in the manufacture of moulds or kernels (Schütz & Wolf, 1980). Phenol concentrations of 1-4 mg/m³ were measured in a foundry in Osaka, Japan (Kuwata et al., 1980). Local phenol concentrations were reported to be as high as 75-420 mg/m³ due to the thermal degradation of the resin. However, this effect of thermal degradation was not reflected in hourly concentrations measured during the foundry process: values of 3-16 mg phenol/m³, with an average of 10 mg phenol/m³, were reported by Schütz & Wolf (1980), and a maximum hourly average of 2.7 mg phenol/m³ was reported by Ryser & Ulmer (1980). (It is not known whether these results were obtained in personal or area air samples). Phenol concentrations during the operation of an electric furnace in a steel factory in Pueblo, Colorado, USA, were 0.04, 0.18 and 0.20 mg/m³ in the vicinity of the furnace. General room air samples taken during operation of a grey iron foundry were below the detection limit (Gunter, 1987).

5.2.3 Other occupational situations

Exposure levels of 5-88 mg phenol/m³ have been reported for employees in the ex-USSR who quenched coke with waste water containing 0.3-0.8 g/litre phenol (Petrov, 1960).

Measurements at coal gasification and liquefaction plants in the USA showed relatively low phenol concentrations (≤ 0.08 mg per m³) at various sites (Dreibelbis & Hawthorne, 1985).

In a Japanese bakelite factory, area samples contained 0-12.5 mg phenol/m³ (Ohtsuji & Ikeda, 1972).

In a synthetic fibre factory in Japan, concentrations of 19 mg phenol/m³ were measured, whereas in a USA fibrous glass wool factory concentrations of 0.05-1.3 mg phenol/m³ were reported (Dement et al., 1973; Ogata et al., 1986).

The concentrations of phenol in creosote vapour, analysed in seven creosote impregnation plants in Finland, ranged from < 0.1 to 1.8 mg/m³ air (Heikkilä et al., 1987). The highest exposures occurred during the cleaning of creosote warming chambers.

During the dissection of cadavers by dental students, phenol breathing zone concentrations ranged from 5 to 19 mg phenol/m³. (The high phenol concentrations resulted from the applied embalming solution, in combination with inadequate ventilation) (Boiano, 1985).

5.3 General population exposure

5.3.1 Indoor air

Borovik & Dmitriev (1981) found a maximal concentration of 0.02 μg phenol/m³ in hospitals in the ex-USSR. It is, however, not clear from where the phenol originated; it may have been used as a disinfectant in these hospitals.

No information has been found with regard to phenol concentrations in residential houses and apartments. Cigarette smoking must be considered as the most important potential source in dwellings. A distinction should be made between the main stream (the smoke inhaled by the smoker) and the side stream (produced by the smouldering cigarette itself). It was estimated that 0.01-0.22 mg phenol per cigarette was released in the mainstream, while the sidestream phenol content was 2.6 times higher. In the case of various Japanese cigarettes, 0.3-0.4 mg phenol was emitted into air during burning (Kuwata et al., 1980). For an unventilated room of 50 m³, the smoking of one cigarette would thus result in a phenol concentration of 6-8 μg/m³.

5.3.2 Food and drinking-water

Phenol is found in smoked meat and fish products. The wood smoke with which such products are treated contains, among other ingredients, a wide range of phenols and phenol ethers, which contribute significantly to the characteristic smoke aroma (smell and taste) of the product.

Phenol is absorbed into the food products during smoking. Quantitative data, however, are scarce, since phenols are usually determined as a group. According to Toth (1982), the total phenol content of smoked sausage is 70 mg/kg; Bratzler et al. (1969) found a content of 37 mg/kg in the outer layer of the product, and lower contents in the inner part. Luten et al. (1979) determined a number of individual phenols in smoked herring and found a phenol content, depending on the duration of smoking, of approximately 10-30 mg/kg. Potthast (1976, 1982) measured 2-18 mg/kg in smoked ham and liver sausage.

If liquid smoke derivates are used in order to give a smoky flavour to fish and meat products, the end product also contains phenol. However, with regard to smell and taste aspects, phenol is not the most important phenolic compound from wood smoke. In this respect, methoxy and dimethoxy phenols are more important, together with aliphatic fatty acids and carboxyl compounds.

Phenol may also enter food unintentionally by, for instance, contamination in transport or from packaging materials, or contact with other phenol-containing materials. However, these accidental cases would probably be detected and lead to non-acceptance by the consumer, owing to the conspicuous phenol smell and taste (see section 8.3 for odour and taste thresholds).

Phenol has been found in botton pits fish from 5 sites in Commencement Bay in Tacoma, USA, at a maximum average of 0.14 mg/kg and an overall maximum of 0.22 mg/kg (Nicola et al., 1987).

Little information is available with respect to the occurrence of phenol in drinking-water. Surface and ground waters intended for the production of drinking-water in the Netherlands were reported to contain 1-9 µg phenol/litre (phenol index, including other phenolic compounds) (RIVM, 1986). Phenol was found in a domestic water supply in the USA at a level of 1 µg/litre (Ramanathan, 1984). Cases of drinking-water pollution with phenol have been reported in the UK and the USA; the phenol water concentrations were reported to be 5-10 µg and 5-120 mg per litre, respectively (see section 8.1.2.2). Chlorination of drinking-water may result in the formation of chlorophenols from phenol, which greatly adds to the objectionable smell and taste (Jarvis et al., 1985).

6. KINETICS AND METABOLISM IN LABORATORY ANIMALS AND HUMANS

Phenol and conjugated metabolites of phenol occur naturally in animal and human tissue and can be detected in the urine, faeces, saliva and sweat. The body's production of phenol depends on the type of diet: a high protein or meat diet promotes phenol formation.

6.1 Absorption

Phenol is readily absorbed through all routes, such as the lungs, intact and abraded skin, and the gastrointestinal tract of both humans and animals (Von Oettingen, 1949; Deichmann & Keplinger, 1963).

6.1.1 Animal uptake studies

6.1.1.1 Pulmonary

There are no *in vivo* data on absorption of phenol following inhalation exposure. However, *in vitro* studies by Hogg et al. (1981), using ^{14}C-phenol with excised trachea-lung preparations and isolated perfused rat lung, demonstrated that phenol can be rapidly and efficiently absorbed in the lungs.

6.1.1.2 Dermal

The extent of absorption of phenol through rabbit skin is more strongly influenced by the area of the skin exposed than by the concentration of the applied solution in water (Deichmann & Witherup, 1944; Liao & Oehme, 1980).

In studies with the hairless mouse, phenol destroyed the stratum corneum (Behl et al., 1983a). Similar effects were reported by Huq et al. (1986) and Jetzer et al. (1986). Absorption of phenol through thermally damaged mouse skin *in vitro* was also reported to be greatly enhanced (Behl et al., 1983b). In contrast, Deichmann et al. (1952) observed that injury of the rabbit skin caused by phenol appeared to retard the rate of absorption.

The permeability of mouse skin to phenol from aqueous solution *in vitro* increased with increasing temperature of the carrier solution (from 10 to 37 °C) (Jetzer et al., 1988).

Measurement of the permeation constant of phenol through hairless mouse skin at 37 °C *in vitro* yielded a value of 18 800 ± 3000 cm/h (Huq et al., 1986; Jetzer et al., 1986).

6.1.1.3 *Intestinal*

When a single oral dose of 25 mg/kg body weight was administered to rats, pigs or sheep, more than 95% was absorbed (Kao et al., 1979).

In vitro studies showed that aqueous solutions of phenol placed into ligated sections of the gastrointestinal tract had the fastest absorption rate in the colon, followed by the ileum. The absorption rate in the stomach was much slower (Deichmann & Keplinger, 1963).

6.1.2 *Human uptake studies*

6.1.2.1 *Pulmonary*

The retention of phenol in the bodies of eight human volunteers exposed to 6-20 mg/m³ by inhalation only for 8 h was 70%-80% during the course of the study (Piotrowski, 1971). Ohtsuji & Ikeda (1972) reported similar observations.

6.1.2.2 *Dermal*

Human skin absorption of phenol vapour (5-25 mg/m³) occurs rapidly (Ruedemann & Deichmann, 1953). Fatal cases reflect the rapid rate of absorption of phenol through the skin (Turtle & Dolan, 1922; Duverneuil & Ravier, 1962; Hinkel & Kintzel, 1968; Lewin & Cleary, 1982). The retention in eight human volunteers, exposed to phenol vapour at concentrations of 6-20 mg/m³, by skin only, for 6 h was 70-80% (Piotrowski, 1971). Piotrowski (1971) proposed the following formula for calculating the absorption rate of phenol vapour through the skin:

$$A = (0.35)C$$

where A is the amount of phenol absorbed in mg/h per unit area and C is the phenol air concentration in mg/m³.

Concentrations of between 5 and 10% phenol denature epidermal protein, and this can partly prevent absorption. The phenol-protein complex is not stable and by dissociation of phenol the substance may exert its action over a period of time (Schmidt & Maibach, 1981).

Phenol was detected in the urine of 4 out of 16 infants (2-5 months) with seborrhoeic eczema who were skin-painted twice daily for 48 h with a commercial paint containing 4% (w/v) phenol and 8% (w/v) resorcinol (Rogers et al., 1978). In adults, a single topical application of 4 μg phenol/cm^2 on 13 cm^2 of the ventral forearm, reportedly gave an absorption of 4.4% of the administered dose (Feldman & Maibach, 1970). The period of exposure and the concentration of phenol are both factors that determine the extent of absorption (Piotrowski, 1971; Roberts et al., 1977; Baranowska-Dutkiewicz, 1981).

In vitro studies have also shown that phenol from aqueous solutions (1% w/v) readily penetrates human skin (Roberts et al., 1977, 1978). A value of 8200 cm/h was obtained as the permeation constant of phenol through human skin at 25 °C (Flynn & Yalkowsky, 1972). In an *in vitro* study with human abdominal skin, 10.9% of the applied dose was absorbed. This study showed an excellent qualitative, but a somewhat less accurate quantitative, agreement between the *in vivo* and *in vitro* skin absorption of 12 compounds (Franz, 1975).

6.2 Distribution

Phenol is rapidly distributed to all tissues in exposed animals.

In rabbits, 15 min after oral administration of 0.5 g phenol/kg, chemical analysis indicated that the liver contained the highest concentration of total phenol followed by the central nervous system, lungs and blood. After 82 min, phenol was fairly uniformly distributed over all tissues. The proportion of free to conjugated phenol changed with time, and, after 360 min, most of the phenol was conjugated (Deichmann, 1944).

After a single oral administration of ^{14}C-phenol (207 mg/kg) to rats, the highest concentration ratios between tissue and plasma were found in liver (42%), followed by spleen, kidney, adrenal, thyroid and lungs, with a peak tissue level occurring after 0.5 h (Liao, 1980; Liao & Oehme, 1981a).

Highest tissue residues were found after 2 h in the kidneys and livers of mice and rats treated intravenously (Gbodi & Oehme, 1978; Wheldrake et al., 1978; Greenlee et al., 1981).

6.3 Metabolic transformation

6.3.1 Metabolite identification

Studies employing several species have demonstrated that conjugation with glucuronic acid and sulfate are major metabolic pathways for phenol. Hydroxylation to hydroquinone and catechol also occurs (Williams, 1938, 1959; Garton & Williams, 1949; Bray et al., 1952a,b,c; Parke & Williams, 1953).

In vitro studies have shown the formation of 4, 4′-biphenol and diphenoquinone by neutrophils, activated leucocytes and by horseradish peroxidase following addition of phenol (Eastmond et al., 1986).

Phenol metabolism in rabbits was studied by Deichmann & Keplinger (1963). During the first 24 h following oral administration of a sublethal dose of 300 mg phenol/kg body weight, 23% of the administered dose was recovered as exhaled carbon dioxide. Trace amounts of catechol and hydroquinone were also detected in the breath. Over the same period, 72% of the dose was excreted in the urine (48% of which was excreted as free and 52% as conjugated phenols), 1% was excreted in the faeces, 4% remained in the carcass, and trace amounts were exhaled.

Oral administration of ^{14}C-phenol (1.2 mg/kg) to rats resulted in at least 80% excretion in urine within 24 h, with 68% as phenyl sulfate and 12% as phenyl glucuronate (Edwards et al., 1986).

A pronounced shift from sulfation to glucuronidation was observed in rats after increasing the phenol dose (Koster et al., 1981). This observed shift is apparently due to a saturation of the overall sulfation process, rather than to a depletion of inorganic sulfate (Weitering et al., 1979; Koster et al., 1981; Koster, 1982). A limited availability of 3-phosphoadenosine-5-phosphosulfate may account for the decreased proportion of phenol conjugation to sulphate at relatively high doses (Ramli & Wheldrake, 1981). Repeated administration of phenol, however, did not affect glucuronide synthesis in rats (Takemori & Glowacki, 1962).

The pig has limited ability for phenol sulfation. The domestic cat lacks the ability for glucuronic acid conjugation of phenol. In cats, phenyl phosphate has been detected as a metabolite in small amounts, in addition to sulfate conjugates (Capel et al., 1974; French et al., 1974).

Following oral administration of ^{14}C-phenol (0.01 mg/kg) to three men, 90% of the dose was excreted in the urine within 24 h, mainly as phenyl sulfate (77%) and phenyl glucuronide. Small amounts of guinol sulfate and guinol glucuronide were also present (Capel et al., 1972b).

Several investigators have confirmed the above-mentioned results using *in vitro* methods (DeMeio & Arnolt, 1944; Capel et al., 1972b; Shirkey et al., 1979; Hogg et al., 1981; Koster et al., 1981; Sawahata & Neal, 1983).

6.3.2 Covalent binding to macromolecules

Early pharmacokinetic studies (measuring distribution volumes) in dogs, pigs and goats suggested that tissue binding occurs (Oehme, 1969). Further animal studies have indicated that phenol and/or its metabolites bind covalently to tissue protein, mainly in the liver (Bolt, 1977; Illing & House, 1980; Jergil et al., 1982; Smart & Zannoni, 1984). Binding to rabbit bone marrow mitochondrial DNA in studies with isolated cells has also been reported (Rushmore et al., 1984). *In vivo* and *in vitro* studies have demonstrated covalent binding of radiolabelled phenol to plasma proteins from humans, dogs, rats and trout (Liao, 1980; Liao & Oehme, 1981a,b; Judis, 1982; Schmieder & Henry, 1988). Reactive phenol metabolites formed by peroxidases bind readily to proteins (Eastmond et al., 1986, 1987a) and DNA (Subrahmanyam & O'Brien, 1985).

6.3.3 Location

Quantitatively, the most important sites of phenol conjugation are the liver, lung and gastrointestinal mucosa. The relative roles played by these tissues depend on the route of administration and the dose.

The liver is an important site of phenol metabolism. After direct administration of phenol into the hepatic circulation, the liver showed considerable first-pass metabolism in rats (Cassidy &

Houston, 1980; Houston & Cassidy, 1982). Phenol-metabolizing enzymes have been detected in rabbit hepatic microsomes (Koop et al., 1989).

Other tissues, such as lungs, intestines and kidneys, also play an important role in phenol metabolism (Quebbemann & Anders, 1973; Powell et al., 1974; Houston & Cassidy, 1982). Phenol sulfotransferases, which catalyse phenol sulfation, occur in a variety of human tissues (intestinal wall, lungs, platelets, adrenal glands, brain, placenta, etc.) (Campbell et al., 1987; Gibb et al., 1987). After oral uptake of phenol, there is a very large first-pass metabolism in the intestines. The lungs also show considerable first-pass metabolism (as was established after direct administration into the pulmonary circulation of rats) (Cassidy & Houston, 1980; Houston & Cassidy, 1982). Due to saturation of hepatic enzymes, extrahepatic tissues play an increasing role in the conjugation of phenol as the dose of phenol increases; at doses higher than 5 mg/kg body weight, intestinal conjugation in rats exceeds the contribution of the hepatic and pulmonary enzymes (Cassidy & Houston, 1984).

Myeloperoxidases isolated from human neutrophils and peroxidative enzymes from activated human leucocytes mediate the formation of reactive phenol metabolites including 4,4′-biphenol and diphenoquinone. Myeloperoxidase-mediated hydroxylation occurs in addition to hepatic cytochrome P-450 oxidation. In several species, myeloperoxidase activity has been reported in bone marrow, where it may play a role in phenol metabolism and toxicity (Eastmond et al., 1986, 1987a; Subrahmanyam et al., 1991).

6.4 Elimination and excretion

Urinary excretion is the major route of phenol elimination in animals and humans. The rate of excretion varies with dose, route of administration and animal species (Deichmann, 1944; Capel et al., 1972a,b). Of 18 animal species studied by Capel et al. (1972a,b), the 24-h urinary excretion of phenol was greatest in the rat (95% of the 25 mg/kg body weight oral dose) and the lowest in the squirrel monkey (only 31% of the dose). Liao & Oehme (1981a,b) reported a half-life of 4 h in rats.

Five days after oral gavage with ^{14}C-phenol (0.1 mg/kg body weight), only 0.3% of the applied dose was retained in rats (Freitag et al., 1985).

Only minor amounts of unchanged phenol are excreted in exhaled air or in faeces (Deichmann & Keplinger, 1963). Less than 1% of an orally administered dose of 300 mg phenol/kg body weight to rabbits was found in the faeces after 24 h (Deichmann, 1944).

Phenol conjugates may also be excreted in the bile of rats (4.6% of a 50 mg/kg dose) (Abou-el-Makarem et al., 1967). It has been suggested that biliary excretion of phenol plays an important role when urinary excretion is impeded. Rats, whose kidneys were ligated, showed a marked increase in biliary excretion of phenol metabolites (Weitering et al., 1979). Furthermore, it has been reported that phenol and its metabolites can undergo enterohepatic circulation in rats (Gbodi & Oehme, 1978).

Urinary excretion of phenol in human volunteers exposed to phenol vapour via inhalation (chamber studies) or skin, occurred with an excretion rate constant of $k = 0.2/h$. For a one-compartment model, this corresponds to a half-life of approximately 3.5 h (Piotrowski, 1971).

6.5 Biological monitoring

The US ACGIH has listed a biological exposure index for phenol of 250 mg/g creatinine for end-of-shift urine samples (ACGIH, 1991).

The excretion of phenol and phenol conjugates in the urine may be used as an index of exposure, but it should be noted that there are other causes that may lead to phenol excretion in the urine. One of these is benzene exposure; other possible significant sources are food and drugs (Docter & Zielhuis, 1967; Ikeda & Ohtsuji, 1969; Fishbeck et al., 1975; Paradowski et al., 1981). Elevated urinary phenol excretion is thus not a specific index of exposure to phenol. Furthermore, the large range of "normal" urine values (phenol concentrations have been found to vary from 0.5 to 81.5 mg/litre) (Deichmann & Schafer, 1942; Docter & Zielhuis, 1967; Piotrowski, 1971; Gspan et al., 1984; Pekari et al., 1992) would appear to limit the usefulness of urinary phenol excretion as an accurate index of low occupational exposure levels.

In volunteers, after a single 8-h exposure to phenol vapour concentrations of up to 6.8 mg/m^3, the phenol excretion in urine increased up to a maximum of 100 mg total phenol/litre (Piotrowski, 1971). In workers occupationally exposed to 10 mg phenol/m^3, concentrations in urine of up to 262 mg/litre were reported (Ohtsuji & Ikeda, 1972). However, another recent study, using more specific methods of analysis, showed good correlation (R=0.91) between exposure levels in the range 5-17 mg/m^3 and the total concentration of phenyl sulfate and phenyl glucuronide in the urine at the end of the workshift (Ogata et al., 1986).

7. EFFECTS ON LABORATORY MAMMALS, AND *IN VITRO* TEST SYSTEMS

7.1 Single exposure

7.1.1 LD_{50} values

After oral administration of phenol to mice, rats and rabbits, LD_{50} values ranged from 300-600 mg phenol/kg body weight. No LC_{50} values have been reported in the published literature. However, after inhalation of 900 mg phenol/m^3 by rats for 8 h, no deaths were observed. The dermal LD_{50} (by occlusive and non-occlusive techniques) was 670 mg/kg body weight for rats and 850-1400 mg phenol/kg body weight for rabbits. LD_{50} values for intraperitoneal injection were in the range of 127-223 mg phenol/kg body weight for rats.

A summary of LD_{50} values is given in Table 5.

7.1.2 Effects

The acute lethality of phenol, associated with exposure to high concentrations, is generally attributed to a depressing effect on the central nervous system (see also section 7.8.1). The clinical effects of phenol poisoning are independent of the route of administration. Reported symptoms include neuromuscular hyperexcitability, including twitching and severe convulsions. Heart rate at first increases, then becomes slow and irregular. Blood pressure at first increases slightly, then falls markedly. Salivation, marked dyspnoea and a decrease in body temperature are also among the effects reported (Deichmann & Witherup, 1944; Von Oettingen & Sharples, 1946; Farquharson et al., 1958; Ernst et al., 1961; Deichmann & Keplinger, 1963; Oehme & Davis, 1970; Pullin et al., 1978; Liao & Oehme, 1980; Reid et al., 1982).

After oral ingestion, the mucous membranes of the throat and oesophagus showed swelling, corrosion, and necrosis, with haemorrhages (Deichmann & Keplinger, 1963).

In a study by Schlicht et al. (1992), female Fischer-344 rats were administered 0, 12, 40, 120 or 224 mg phenol/kg body weight by gavage in a water vehicle. Animals were examined for clinical signs, and neurotoxicity and systemic (liver, kidney, adrenal and thymus) effects, 4-20 h after treatment. Tremors

Table 5. Acute animal toxicity of phenol LD_{50} values

Species	Route of administration	LD_{50} values (mg/kg body weight)	Vehicle	Reference
Mouse	oral	300		Von Oettingen & Sharples (1946)
Mouse	oral	427		Kostovetskii & Zholdakova (1971)
Rat	oral	340-530	2-7% in water	Deichmann & Witherup (1944)
Rat	oral	512		Kostovetskii & Zholdakova (1971)
Rat	oral	445-520	water	Thompson & Gibson (1984)
Rat	oral	400	water	Schlicht et al. (1992)
Rat	dermal	670 (570-780)	undiluted	Conning & Hayes (1970); Brown et al. (1975)
Rat	intra-peritoneal	127-223	water or undiluted	Thompson & Gibson (1984)
Rabbit	oral	400-600	2-7% in water	Deichmann & Witherup (1944)
Rabbit	dermal	850 (600-1200)		Flickinger (1976)
Rabbit	dermal	1400 (740-2670)		Vernot et al. (1977)

were observed 1-2 min after dosing in the two highest dose groups. The pupil response to light (miosis) was significantly inhibited at all dose levels at 24 h after exposure. Locomotor activity was reduced at 224 mg/kg. At this dose level, 2/6 animals had hepatocyte necrosis, 4/6 had renal vascular stasis and 4/6 had necrosis of the thymus. At 120 mg/kg, liver necrosis was present in 1/7 animals, as was necrosis of the thymus gland.

In various animal species, inhalation of phenol adversely affected the lungs, causing hyperaemia, infarcts, bronchopneumonia, purulent bronchitis and hyperplasia of the peribronchial tissues (Von Oettingen, 1949).

Sensory irritation was measured in mice by the Alarie assay. A 50% decrease in respiratory rate (RD_{50}) was found at 638 mg phenol/m³ (De Ceaurriz et al., 1981).

Ocular and nasal irritation, tremors and incoordination were reported in rats exposed via inhalation to 906 mg/m³ for 8 h (Flickinger, 1976).

Other pathological abnormalities induced by phenol by various routes of administration included demyelination of nerve fibres (see also section 7.8.1), myocardial degeneration and necrosis (Deichmann & Keplinger, 1963; Liao & Oehme, 1980). Kidney damage (vacuolization and enlargement of cells) and liver damage (e.g., enlargement of hepatic cells) were also observed (Oehme & Davis, 1970; Coan et al., 1982). Urine was usually dark or "smoky" in appearance, probably due to oxidation products of phenol (Solliman, 1957).

7.2 Short-term exposure

7.2.1 Oral exposure

In a study by Schlicht et al. (1992), groups of eight female Fischer-344 rats received oral doses of phenol in a water vehicle of 0, 4, 12, 40 or 120 mg/kg body weight daily for 14 days. Tremors were apparent only after the first dose at the highest level. Exposure to 120 mg/kg per day was lethal to all rats within 11 days. The pupil response (miosis) was decreased one day after the last dose for all but the highest surviving dose group (the incidences were 100%, 50%, 62% and 76% for the 0, 4, 12 and 40 mg/kg groups, respectively). Locomotor activity was not affected after the 4th, 9th or 14th dose. No hepatic effects were observed at 40 mg/kg per day, while 3/8 animals had renal vascular stasis. There were no histological effects at 12 mg/kg per day. At 40 mg/kg per day, the pathological changes in the kidneys included two animals with tubular degeneration in the papillar region, and one with protein casts in the tubules. The pathological report attributed these findings to decreased vascular perfusion (MacPhail, personal communication to the IPCS).

Rats were administered, by gavage, 20 daily doses of 10, 50 or 100 mg phenol/kg body weight. At necropsy, slight effects on liver and kidneys were reported at 100 mg phenol/kg body weight (Dow Chemical Company, 1976).

Rats receiving 50 or 100 mg phenol/kg body weight, by gavage, over a 6-month period (135 doses, presumably daily, 5 days/week) were reported to show slight to moderate kidney damage. Administration of 100 mg phenol/kg body weight apparently resulted in slight liver changes (Dow Chemical Company, 1976).

In a range-finding study, carried out prior to a long-term carcinogenicity study, mice and rats were provided with tap water containing 0, 100, 300, 1000, 3000 or 10 000 mg phenol/litre for 13 weeks. Mean body weight gain was decreased only in mice and rats receiving 10 000 mg phenol/litre (NCI, 1980). In these drinking-water studies, the highest daily doses were calculated to be approximately 2000 mg phenol/kg body weight for mice and 1000 mg phenol/kg body weight for rats.

Phenol was provided to rats in drinking-water for 12 months at 0, 800, 1200, 1600, 2000 and 2400 mg phenol/litre. Depressed weight gain was observed in rats receiving doses $\geq$ 2000 mg/litre. The corresponding daily dose was calculated by the authors of the study to be $\geq$ 200 mg/kg body weight (Deichmann & Oesper, 1940).

7.2.2 Dermal exposure

In a study by Deichmann et al. (1950), rabbits were exposed to 1.18-7.12% phenol in water (64-380 mg phenol/kg body weight) for 5 h/day, 5 days/week, for 18 days. Dose-related systemic effects (tremors, death) were observed in rabbits exposed to $\geq$ 2.37% phenol (130 mg phenol/kg body weight), while skin irritation (hyperaemia, tissue necrosis) occurred at doses of $\geq$ 3.56% phenol (190 mg phenol/kg body weight). This effect was particularly apparent when the application sites were bandaged.

7.2.3 Inhalation exposure

No studies reported or conducted according to contemporary standards were available.

In a study by Deichmann et al. (1944), rats, rabbits and guinea-pigs were exposed to concentrations of 100-200 mg phenol vapour/m^3, 7 h/day, for 5 days/week. Rats exposed for a period of 74 days did not show any gross or microscopic evidence of injury. Rabbits survived a 3-month exposure but, at autopsy, lung and heart damage and indications of liver and kidney damage were found. Guinea-pigs were the most susceptible. Five out of twelve died after 12 days of exposure, and the remaining seven were killed after 29 days of exposure. Prior to death, guinea-pigs showed weight loss, respiratory difficulties, and signs of paralysis. At autopsy, there was evidence of acute lobular pneumonia, vascular damage, and hepatic and renal damage; the total (free and conjugated) phenol content of the blood was 14 mg/litre. The rabbits had similar, but less severe, symptoms.

Groups of 10 monkeys, 50 rats and 100 mice were exposed to 19 mg phenol/m^3, 8 h/day, 5 days/week for 90 days. Concurrent control groups were exposed to fresh air only. No deaths occurred and there was no reduction in weight gain of treated animals. There were no statistically significant adverse effects observed when the animals were assessed by a stress test involving swimming performance. A range of clinical chemistry, haematology and urinalysis parameters were not affected by exposure to phenol. Routine histology was performed on the liver, lungs, kidneys, brain and heart. The results of the percentage of animals showing evidence of "pathological change" indicated effects in the liver and kidneys of exposed animals. However, the author of the study concluded that no clinical or pathological changes occurred that were of toxicological importance. It is not clear if the upper respiratory tract was examined in this study in order to look for evidence of irritation (Sandage, 1961).

Continuous exposure to 100 mg phenol/m^3 for 15 days significantly affected the central nervous system of rats, as was demonstrated by their performance in the "tilted plane" test. Plasma levels of potassium, magnesium, lactate dehydrogenase, aspartate aminotransferase (ASAT), alanine aminotransferase (ALAT) and glutamate dehydrogenase were elevated. Haemoglobin, haematocrit, and plasma sodium, calcium and chloride levels were unaffected (Dalin & Kristofferson, 1974).

7.2.4 Subcutaneous exposure

Subcutaneous exposure to phenol was studied principally to obtain information about neurological or haematopoietic effects (see sections 7.8.1 and 7.8.2). No other effects were reported.

7.2.5 *Ear exposure*

Instillation of phenol (form and amount not specified) into the inner ear round window of Sprague-Dawley rats caused morphological damage to the organ of Corti in the basal coil. The outer hair cells appeared to be more sensitive to phenol than the inner hair cells, which were mostly intact. As a result of the damage, impairment of inner ear function was noted (as determined by auditory brain stem recordings) which was regressive for lower sound frequencies, but appeared to be permanent for higher frequencies (Anniko et al., 1988).

7.3 Skin and eye irritation; sensitization

Local damage to the skin, following exposure to phenol, was found to include erythema, inflammation, discoloration, eczema, papillomas and necrosis (Deichmann, 1949; Deichmann et al., 1950; Conning & Hayes, 1970; Pullin et al., 1978). For example, in rabbits, 0.5 g phenol, moistened with physiological saline, produced necrosis of both the intact and abraded skin (Flickinger, 1976).

Solutions of 10-14% (v/v) phenol in water have been reported to cause transient delayed erythema (after 0.5-5 h) and acute vascular permeability, as assessed by exudation of intravenously injected Evans blue, in guinea-pigs after dermal treatment for 1 min (Steele & Wilhelm, 1966).

In one study, an increase in ear thickness was used as an index of skin irritation (inflammation). Maximal responses to phenol were observed one hour after application of 1-2 mg phenol to the ear of female ICR mice. Significant thickening could still be detected 6 weeks after exposure (Patrick et al., 1985).

When phenol, in glycerine dilutions down to 10% or 5% aqueous solutions, was applied to the rabbit eye, severe damage (complete destruction to opaque corneas) was seen. Immediate water irrigation was very effective in preventing the opacity. A delay of 10 seconds reduced this effectiveness (Murphy et al., 1982).

Fourteen days after the application of 0.1 g phenol to the rabbit eye, all eyes exhibited keratoconus and pannus formation (Flickinger, 1976).

Phenol gave negative results in a Magnussen and Kligman skin sensitization test (Itoh, 1982).

7.4 Long-term exposure

No adequate data are available. Studies on carcinogenicity are presented in section 7.7.

7.5 Reproduction, embryotoxicity and teratogenicity

7.5.1 Reproductive toxicity

No adequate studies conducted according to current protocols are available.

Heller & Pursell (1938) exposed rats to 100-12 000 mg phenol/litre drinking-water, corresponding to calculated approximate daily oral doses of 10-1200 mg phenol/kg body weight. General appearance, growth and fecundity were normal for rats exposed to 100-1000 mg/litre for five generations and to 3000 or 5000 mg/litre for three generations. Stunted growth was noted in the offspring of rats exposed to 7000 mg/litre. Many of the offspring died at levels of 8000 mg/litre because of maternal neglect. At 10 000 mg/litre, the offspring died at birth, and at 12 000 mg/litre there was no reproduction.

7.5.2 Embryotoxicity/teratogenicity

7.5.2.1 In vivo studies

Phenol was evaluated for maternal and developmental toxicity in timed-pregnant Sprague-Dawley rats (20-22 confirmed pregnancies per group). Distilled water (vehicle) or phenol (30, 60 or 120 mg/kg per day) was administered daily by gavage in a volume of 5 ml/kg of body weight throughout the period of major organogenesis (gestational days 6-15). Dams were weighed on the day of sperm detection (gestational day 0), prior to daily dosing, and at termination (gestational day 20); observations for clinical signs of toxicity were conducted during the treatment period. At termination (gestational day 20), maternal liver weight, gravid uterine weight and status of uterine implantation sites (i.e. number of implants, resorptions, late fetal deaths and live fetuses) for each dam were recorded. Each live fetus was weighed, sexed and examined for external morphological abnormalities. Visceral examination of each fetus was performed using a fresh tissue dissection method; approximately one-half of the fetal heads from each litter were fixed (Bouins' solution) and sectioned free-hand for examination of internal structures; carcasses (one-half without heads) were cleared and stained with Alizarin Red S prior to skeletal examination. All control and phenol-treated dams survived to scheduled sacrifice, and no distinctive treatment-related signs of toxicity were noted. Pregnancy rates at termination were high (95-100% per group) and no litters were totally resorbed, so that a total of 20-22 live litters per group (268-293 fetus per group) was available for examination. No significant dose-related changes were noted for the following end-points: maternal body weight (gestational day 0, 6, 11, 15 or 20), maternal body weight gain (treatment period, gestational period or gestational period corrected for gravid uterine weight), maternal liver weight, gravid uterine weight, prenatal mortality, live litter size or incidence of morphological abnormalities (malformations or variations). However, average fetal body weight per litter was significantly reduced at the high-dose (93% of average control weight) (Jones-Price et al., 1983a).

In a study by Kavlock (1990), phenol was administered by oral gavage to groups of Sprague-Dawley rats on day 11 of gestation (day 1 : sperm plug) at 0, 100, 333, 667 and 1000 mg phenol/kg body weight. The vehicle used in this study was a 4:4:1:1 mixture of water, Tween 20, propylene glycol and ethanol. Maternal toxicity (decreased weight gain) was seen at the two highest doses. Offspring viability and growth were not affected up to postnatal day 6,but hind limb paralysis was observed in some offspring in the two highest dose groups.

In a screening assay, groups of 17-21 Fischer-344 rats received 0, 40 or 53.3 mg phenol/kg body weight by gavage in water on gestation days 6-15. There were no significant effects on maternal body weight gain. One of 15 pregnant females resorbed the entire litter at 40 mg/kg and 2 of 16 did so at 53.3 mg/kg (there were no similar effects in 153 control litters in the study). All three females had severe respiratory syndromes (rales and dyspnoea). One high-dose female with symptoms of respiratory toxicity delivered a low weight litter that had poor viability. Kinked tails were present in 2 of 4 surviving pups in that litter. Litter size on postnatal days 1 and 6 was significantly reduced at 53.3 mg/kg but not at 40 mg/kg. There were no effects on pup body weights on postnatal days 1 or 6 (Narotsky & Kavlock, 1993).

Phenol was evaluated for maternal and developmental toxicity in timed-pregnant Swiss albino (CD-1) mice (22-29 confirmed pregnancies per group). Distilled water (vehicle) or phenol (70, 140 or 280 mg/kg per day) was administered daily by gavage in a volume of 10 ml/kg of body weight throughout the period of major organogenesis (gestational days 6-15). Dams were weighed on the day of vaginal plug detection (gestational day 0), prior to daily dosing (gestational days 6-15), and at termination (gestational day 17); observations for clinical signs of toxicity were conducted during the treatment period. Evaluation of maternal and developmental end-points at termination (gestational day 17) were the same as for rats (see description from the study by Jones-Price et al., 1983a, above). Toxicity observed at the high-dose level included 11% mortality (4/36 treated females), clinical signs (especially tremor and ataxia), reduced maternal body weight (gestational day 17), reduced maternal body weight gain (treatment period, gestational period and gestational period corrected for gravid uterine weight), and a trend only toward reduced maternal liver weight. Pregnancy rates at termination ranged from 71 to 84%; no litters were totally resorbed so that 22-29 live litters (214-308 fetuses) were available for examination. No dose-related changes were noted for prenatal mortality, live litter size or incidence of morphological abnormalities, except for an apparent increase in cleft palate (8/214 fetuses in the high dose versus 0/308 among controls). (It should be noted that cleft palate is a malformation to which the CD-1 mouse is predisposed under conditions of maternal stress). Average fetal body weight per litter was significantly reduced (82% of average control weight) in the highest dose group (Jones-Price et al., 1983b).

In a study by Minor & Becker (1971), groups of Sprague-Dawley rats were given 20, 63, or 200 mg phenol/kg body weight intraperitoneally on days 9-11 or 12-14 of gestation. Fetal body weight was reduced in the highest dose group treated on days 12-14. No gross anomalies were observed, and intrauterine death was not increased at any dose level.

7.5.2.2 In vitro *studies*

In the chick embryotoxicity screening test (CHEST), 130 substances were tested. For each compound, 120 selected White Leghorn Fowl embryos, aged 1.5, 2, 3 and 4 days of incubation, were used. Phenol did not exhibit embryotoxic properties in this test up to 100 μg, and was one of the least embryotoxic compounds tested (Jelinek et al., 1985).

In a study by Oglesby et al. (1992), phenol was added to cultures of five rat embryos on gestational day 10 at concentrations of 0 to 100 μg/ml. Embryos were examined 42 h later for viability, growth and morphology. Viability was not affected at any concentration, but a low incidence of tail defects was observed at 100 μg/ml, and embryonic growth was decreased at 75 and 100 μg/ml. When hepatocytes isolated from pregnant rats were co-cultured with the embryos, the toxicity to the embryos was increased. Tail defects were observed at 25 and 50 μg/ml, and growth was reduced at these concentrations. Without the presence of hepatocytes, phenol was the least toxic of 13 para-substituted phenols tested in this system; however, it was the only one which became more embryotoxic when hepatocytes were present.

When phenol was added to cultures of human embryonic palatal mesenchyme cells, cell growth was 50% inhibited at a concentration (IC_{50}) of 0.8 mM (78 μg/litre) (Pratt & Willis, 1985).

7.6 Mutagenicity and related end-points

Data on mutagenicity and related end-points are summarized in Tables 6, 7, 8 and 9, respectively.

7.6.1 *Mutagenicity studies*

7.6.1.1 *Bacterial systems*

Phenol has been tested for mutagenicity by a number of authors in various strains of *Salmonella typhimurium* and was shown to be reproducibly negative, both with and without metabolic activation (Epler et al., 1979; Gilbert et al., 1980; Rapson et al., 1980; Pool & Lin, 1982; Haworth et al., 1983). A positive effect was observed in strain TA98 in the presence of an exogenous metabolic activation system in a study employing a modified culture medium (Wild et al., 1980; Gocke et al., 1981).

A positive effect was reported for phenol in a fluctuation test with strain TA100 after metabolic activation, but no data were given on toxicity (Koike et al., 1988; abstract). A positive result was reported in a mutation test with *Escherichia coli* B/Sd-4;

Table 6. Tests for genotoxicity in bacteria

Species	Strain	Measured end-point	Test conditions	Metabolic activation[a]	Results[b]	Reference
Escherichia coli	Sol-4	reverse mutation	0.1-0.2%; 3-24 h (survival less than 2%)	-	+	Demerec et al. (1951)
	AB1899 nm	filamentation	10-500 µg/ml; 3-4 h		-	Nagel et al. (1982)
Salmonella typhimurium	TA100	reverse mutation	fluctuation test 0-500 ng/well	- + (rat)	- + (no data on toxicity)	Koike et al. (1988)
	TA98 TA100	reverse mutation	1000-fold concentration range in DMSO[c]	+ and -	-	Epler et al. (1979)
	TA1535	reverse mutation	0-100 µg/plate	-	-	Gilbert et al. (1980)
	TA1538	reverse mutation	0-50 µg/plate	-	-	
	TA98 TA100 TA1535 TA1537	reverse mutation	0-3333 µg/plate in DMSO; preincubation	- + (rat) + (hamster)	- - -	Haworth et al. (1983)

Table 6 (contd).

Strains	Test	Dose			Reference
TA98 TA100 TA1535 TA1537	reverse mutation	0-3333 μg/plate in preincubation H_2O;	- + (rat) + (hamster)	- - -	Haworth et al. (1983)
TA98 TA100 TA1535 TA1537 TA1538	reverse mutation	0.5-5000 μg/plate in DMSO (5000 μg toxic)	+ and - (rat)	-	Pool & Lin (1982)
TA100	reverse mutation	0.1-1000 μg/plate	-	-	Rapson et al. (1980)
TA98	reverse mutation	0-100 μmol/plate (99.5% purity with 0.15% cresols as main impurity, non-standard media)	- + (rat)	- +	Wild et al. (1980)

[a] + = present; - = absent
[b] + = positive; - = negative
[c] DMSO = dimethyl sulfoxide

Table 7. Tests for genotoxicity in non-mammalian eukaryotic systems

Species	Strain	Measured end-point	Test conditions	Metabolic activation	Results[a]	Reference
Fungi						
Saccharomyces cerevisiae	D3	mitotic recombination	10^{-5}, 10^{-3} dilution of phenol in saline	– + (rat)	– +	Cotruvo et al. (1977)
Aspergillus nidulans		mitotic segregation	5-20 mM	–	+	Crebelli et al. (1987)
Insects						
Drosophila melanogaster	Oregon-R	SLRL[b]	phenol vapour 24 h; 0.2, 0.25, 0.5% in saline, injection		– –	Sturtevant (1952)
			0.01, 0.1, 1.2% in Holtfreter solution; vaginal douch		–	

Table 7 (contd).

Species	Strain	Measured end-point	Test conditions	Metabolic activation	Results[a]	Reference
	Berlin K	SLRL	50 nM in 5% saccharose; feeding, 3 broods F_1 generation		-	Wild et al. (1980); Gocke et al. (1981)
			injection		-	Woodruff et al. (1985)
Fish						
Salmo gairdneri		chromosomal aberrations	0.3–0.6 µl/litre, 72 h		+	Al-Sabti (1985)

[a] + = positive; - = negative
[b] SLRL = sex-linked recessive lethal mutations

Table 8. *In vitro* phenol genotoxicity in mammalian cells

Species	Cell type	End-point[a]	Conditions	Activation[b]	Result[c]	Reference
Chinese hamster	CHO-WBL	CA	500-800 µg/ml	-	-	Ivett et al. (1989)
			2000-3000 µg/ml	+ (rat)	+	
Chinese hamster	V79 lung	forward mutation HPRT	0-500 µg/ml (500 µg/ml toxic)	+ (mouse)	+	Pashin & Bakhitova (1982)
Chinese hamster	CHO-WBL	SCE	300-400 µg/ml	-	+	Ivett et al. (1989)
			2000-3000 µg/ml	+ (rat)	+	
Chinese hamster	V79 lung	intercellular communication	not reported	-	-	Chen et al. (1984)
	V79 lung	intercellular communication	250 µg/ml	-	-	Malcolm et al. (1985)
	V79 lung	intercellular communication	10-75 µg/ml	-	-	Bohrman et al. (1988)

Table 8 (contd).

Species	Cell	Test	Concentration	Result		Reference
Mouse	L5178Y lymphoma	forward mutation TK	600-1800 µg/ml	+ and -	?	McGregor et al. (1988)
	L5178Y	forward mutation	180-890 µg/ml (530 µg/ml toxic)	-	+	Wangenheim & Bolcsfoldi (1988)
			5.6-41 µg/ml (20 µg/ml toxic)	+ (rat)	+	
Mouse	L5178Y	DNA synthesis inhibition	9.4-940 µg/ml	-	+	Pellack-Walker et al. (1985)
	L5178Y	DNA strand breaks	16-470 µg/ml	-	-	Garberg et al. (1988)
			16-470 µg/ml	+ (rat)	+	
	L5178Y	DNA strand breaks	94 µg/ml	-	-	Pellack-Walker & Blumer (1986)
Human	T-lymphocytes	SCE	0.47-282 µg/ml	-	+	Erexson et al. (1985)
	lymphocytes	SCE	188 µg/ml	-	-	Jansson et al. (1986)
	lymphocytes	SCE	1.7-470 µg/ml (470 µg/ml toxic)	+ (rat)	+	Morimoto & Wolff (1980)

Table 8 (contd).

Species	Cell type	End-point[a]	Conditions	Activation[b]	Result[c]	Reference
	lymphocytes	SCE	282 µg/ml	+ (rat)	+	Morimoto et al. (1983)
Human	fibroblast	DNA repair	0.094-9400 µg/ml		+	Poirier et al. (1975)
	HeLa	DNA synthesis inhibition	188 µg/ml	+ (rat)	+	Painter & Howard (1982)
	WI-38	DNA synthesis inhibition	0.094-9400 µg/ml		+	Poirier et al. (1975)

[a] CA = chromosome aberrations; HPRT = hypoxanthine guanine phosphoribosyl transferase locus; TK = thymidine kinase locus; SCE = sister chromatid exchange

[b] - = absent; + = present;

[c] - = negative; + = positive

Table 9. Phenol genotoxicity in *in vivo* mammalian systems

Species/Strain	Measured end-point	Test conditions (sampling times)	Remarks	Results[a]	Reference
Mouse/CD-1	micronuclei in bone marrow	265 mg/kg, oral (0, 18, 24, 42 or 48 h)	bone marrow depression	+	Ciranni et al. (1988a)
Mouse/CD-1	micronuclei in maternal bone marrow and fetal liver	gestation day 13, 265 mg/kg, oral (15, 18, 24, 30, 36 or 40 h)	maternal bone marrow depression	+	Ciranni et al. (1988a)
Mouse/CD-1	micronuclei in bone marrow	250 mg/kg, oral (30 h)	convulsive seizures	-	Gad-El Karim et al. (1986)
Mouse/CD-1	micronuclei in bone marrow	265 mg/kg, i.p. (18, 24, 42 or 48 h)	bone marrow depression	+	Ciranni et al. (1988a)
Mouse/CD-1	micronuclei in bone marrow	40, 80 or 160 mg/kg, i.p. (18 h)	no bone marrow depression	-	Barale et al. (1990)
Mouse/NMRI	micronuclei in bone marrow	47, 94 or 188 mg/kg, i.p. 0, 24 h (30 h)	no information on toxicity	-	Gocke et al. (1981)

Table 9 (contd).

Species/Strain	Measured end-point	Test conditions (sampling times)	Remarks	Results[a]	Reference
Mouse/Porton	chromosomal aberrations in spermatogonia, primary spermatocytes	2 ml of 0.08, 0.8 or 8 mg/litre solution, oral, daily for five generations		+	Bulsiewicz (1977)
Rat/Sprague-Dawley	chromosomal aberrations in bone marrow	72-180 mg/kg, i.p. 300-510 mg/kg, oral (20 h)	LD^1-LD^{30}, no change in mitotic index	-	Thompson & Gibson (1984)
Rat/Sprague-Dawley	DNA strand breaks (alkaline elution in rat testis)	7.9, 26 or 79 mg/kg, i.p. (2.6 or 24 h) 4, 13.2 or 39.5 mg/kg, i.p. for 5 days		-	Skare & Schrotel (1984)

[a] + = positive; - = negative; i.p. = intraperitoneal injection

however, the applied dose levels were highly toxic (Demerec et al., 1951).

7.6.1.2 *Non-mammalian eukaryotic systems*

Negative data were obtained in the absence of exogenous metabolic activation with the eukaryotic microorganism *Saccharomyces cerevisiae.* At high doses and in the presence of an activation system, a positive result was obtained (Cotruvo et al., 1977). Phenol induced mitotic segregation in *Aspergillus nidulans* (Crebelli et al., 1987).

In *Drosophila melanogaster*, no statistically significant sex-linked recessive lethals were obtained after exposure to phenol via a variety of techniques (Sturtevant, 1952; Wild et al., 1980; Gocke et al., 1981; Woodruff et al., 1985). However, when an unusual technique was used, i.e. exposing isolated gonads *in vitro* and implanting them in host larvae, several types of mutations were induced (Hadorn & Niggli, 1946).

When rainbow trout *(Salmo gairdneri)* were exposed to phenol for 72 h at concentrations of 0.3 and 0.6 μl phenol/litre water, the percentage of chromosomal aberrations in gill and kidney tissue was significantly increased, at both concentrations, in a dose-related way. Of these aberrations, 30% were structural, 45% consisted of aneuploidy, and 25% were non-specified metaphases (Al-Sabti, 1985).

7.6.1.3 *Mammalian* in vitro *systems*

Data on *in vitro* genotoxicity in mammalian cells are given in Table 8.

In a Chinese hamster V79 lung cell/HPRT mutation test, phenol gave a positive result with metabolic activation. The highest dose decreased survival by approximately 50% (Pashin & Bakhitova, 1982). In a mouse lymphoma L5178Y cell/TK mutation test, there were statistically significant and dose-related increases in mutation frequency in the presence and absence of metabolic activation (Wangenheim & Bolcsfoldi, 1988). However, in another laboratory this test yielded non-conclusive results (McGregor et al., 1988).

As part of the US NTP testing program, phenol was evaluated for induction of chromosomal aberrations and sister chromatid exchange (SCE) in Chinese hamster ovary (CHO) cells (Ivett et al., 1989). At a delayed harvest time (22.5 h), there were significantly increased incidences of aberrations in cultures that included a metabolic activation system from induced rat liver. Although a dose-response effect was seen, the frequency of aberrations in the absence of activation was low and the authors reported a negative result. Regarding SCE induction, positive results were obtained, both with and without activation. In additional studies, phenol induced SCE in human lymphocytes *in vitro*, both in the presence and in the absence of metabolic activation (Morimoto & Wolff, 1980; Morimoto et al., 1983; Erexson et al., 1985). Negative results (SCE) have also been reported (Jansson et al., 1986).

Phenol gave negative results in three studies in Chinese hamster V79 cell metabolic cooperation assays (Chen et al., 1984; Malcolm et al., 1985; Bohrman et al., 1988).

7.6.1.4 *Mammalian* in vivo *system: somatic cells*

In a bone marrow micronucleus test, groups of four Swiss CD-1 mice (sex not specified) were orally administered 265 mg phenol/kg body weight and were sacrificed at 0, 18, 24, 42 and 48 h. Bone marrow depression (decreased polychromatic erythrocytes/normocytes (PCE/NCE) ratio) persisted at least up to 48 h after dosing. A slight, but statistically significant, increase in the number of micronuclei was seen at 24 h (3 micronuclei/1000 cells versus 1.5 micronuclei/1000 cells; 3000 cells scored per mouse) (Ciranni et al., 1988b).

In a further study to asses the transplacental clastogenicity of phenol, groups of 4 pregnant Swiss CD-1 mice received 265 mg phenol/kg/body weight by oral gavage on day 13 of gestation. After 15, 18, 24, 30, 36 or 40 h, animals were sacrificed and adult bone marrow cells and fetal liver cells were scored for micronuclei. Slight, but statistically significant, increases in the frequency of micronucleated PCE in adult bone marrow were observed at 15, 18 and 24 h (3.8, 4.0 and 5.0 micronuclei/1000 cells, respectively, compared with 2/1000 for negative controls). A statistically significant reduction in the PCE/NCE ratio was seen at 18 and 36 h. Phenol had no effect on the frequency of micronuclei in fetal liver (Ciranni et al., 1988a).

Bone marrow liver cells were also evaluated for the formation of micronuclei, 30 h after oral administration of 0 or 250 mg phenol/kg body weight to groups of five males Swiss CD-1 mice (Gad-El Karim et al., 1986). Uptake was indicated by convulsive seizures in all mice receiving phenol (1000 PCEs scored per mouse).

Phenol (265 mg/kg body weight), administered to Swiss CD-1 mice by a single intraperitoneal injection, was reported to increase the frequency of micronuclei in bone marrow PCEs 18 h post-treatment (7 micronuclei/1000 cells). The increased frequency decreased at 24 h and was no longer statistically significant at 42 h. A decreased PCE/NCE ratio persisted in tests up to 48 h post-treatment (Ciranni et al., 1988b).

Barale et al. (1990) reported a negative result in a bone marrow micronucleus test in Swiss CD-1 mice 18 h after treatment with phenol. There was no effect on the PCE/NCE ratio.

Gocke et al. (1981) briefly reported a negative result in a bone marrow micronucleus test, in which NMRI mice were sampled at 30 h, following i.p. injection of 47-188 mg phenol/kg. No information on toxicity was given.

The results of these and other studies are summarized in Table 9.

7.6.1.5 Mammalian in vivo *systems: germ cells*

Skare & Schrotel (1984) obtained negative results in studies of DNA strand breakage in rat testis. In one experiment, rats received by intraperitoneal injection 0, 7.9, 26 or 79 mg phenol/kg body weight and were sacrificed at 2.6 or 24 h post-treatment. Similar results were obtained with further groups of rats that received 4, 13.2 and 39.5 mg phenol/kg body weight daily for 5 days before sacrifice.

In an unconventional study involving dosing (0, 6.4, 64 and 640 mg phenol/kg body weight) of five successive generations of male and female mice, large numbers of structural and numerical chromosomal aberrations were reported in spermatocytes and spermatozoa, with dose- and generation-related increases. The study was carried out with 138 male mice from an inbred stock after skin testing (Bulsiewicz, 1977).

7.7 Carcinogenicity

The evidence for the carcinogenicity of phenol in experimental animals, based on the studies summarized below, was recently considered by the IARC (1989) to be inadequate. Phenol was classified by US EPA in Group D (data inadequate for evaluating the carcinogenic potential) (Bruce et al., 1987).

7.7.1 Oral exposure

In an NCI (1980) study, groups of 50 male and 50 female B6C3F$_1$ mice were given drinking-water containing 0, 2500 or 5000 mg phenol/litre for 103 weeks. As matched controls, groups of 50 male and 50 female mice received tap water. There was a dose-related decrease in water consumption and mean body weight gain in all groups of mice. In mice receiving 5000 mg phenol/litre, an increase in the number of uterine endometrial stromal polyps (5/48 = 10%) was observed (in matched controls the incidence was 1/50 = 2%). There was no evidence of an increased incidence of malignant tumours. The other observed neoplasms were of the usual number and type found in mice of this strain and age (NCI, 1980).

Groups of 50 male and 50 female Fischer-344 rats received 0, 2500 or 5000 mg phenol/litre drinking-water for 103 weeks, while the matched control group received tap water. Male and female rats given 5000 mg/litre showed a decrease in mean body weight from week 20 onwards. There were statistically significant increased incidences of phaeochromocytomas, leukaemias, lymphomas and C-cell thyroid carcinomas in males of the low-dose group (NCI, 1980). NTP considered this study negative for carcinogenicity due to the lack of a dose-response for the neoplasms and the lack of response in females.

7.7.2 Dermal exposure

Three studies examined the potential carcinogenicity of phenol following dermal application (Rusch et al., 1955, Boutwell et al., 1956; Bernard & Salt, 1982). However, none is considered adequate for the evaluation of carcinogenicity due to the short duration of exposure and/or use of inappropriate vehicles.

7.7.3 Inhalation exposure

No studies have been reported for this route of exposure.

7.7.4 Two-stage carcinogenicity studies

A dose of 3 mg phenol in acetone was applied to ICR/Ha Swiss mice 3 times per week for 52 weeks after initiation with 150 µg DMBA. Papilloma development was enhanced, compared with that of mice exposed to DMBA alone (Van Duuren et al., 1968; Van Duuren & Goldschmidt, 1976). These observations were in agreement with those from earlier reports on the promotional activity of phenol (Boutwell et al., 1955, 1956; Salamon & Glendenning, 1957; Boutwell & Bosch, 1959; Wynder & Hoffmann, 1961).

A dose of 3 mg phenol in acetone applied 3 times weekly for 460 days to female ICR/Ha Swiss mice after initiation with 5 µg benzo[a]pyrene had a slight promoting activity. Simultaneous application of both agents showed a partial reduction in carcinomas compared with mice treated with benzo[a]pyrene alone (Van Duuren et al., 1971, 1973; Van Duuren & Goldschmidt, 1976).

7.8 Special studies

7.8.1 Neurotoxicity

Tremors, convulsions, coma and death were reported after intraperitoneal and subcutaneous doses of phenol (Deichmann & Witherup, 1944; Ernst et al., 1961; Windus-Podehl et al., 1983). The tremors were enhanced by prior monoamine depletion following reserpine treatment (Suzuki & Kisara, 1985).

Upon histological examination of rats in which convulsions had been induced by subcutaneous phenol injection (200 mg/kg body weight given once a week for two weeks), two out of six animals showed spinal cord and spinal root degeneration (Veronesi et al., 1986).

Phenol, given either intravenously or intra-arterially (250 µg), facilitated neuromuscular transmission and antagonized neuromuscular blockade by D-tubocurarine in cats. The effect was determined to be pre-synaptic in origin (Blaber & Gallagher, 1971).

Phenol caused diminution of the compound action potential in preparations of the saphenous nerve after acute and chronic perfusion (Schaumburg et al., 1970).

Groups of five male CD-1 mice were supplied with drinking-water containing 0, 4.7, 19.5 or 95.2 mg phenol/litre for 4 weeks, at the end of which the concentrations of various neurotransmitters and their metabolites were measured in different parts of the brain. The largest effects were seen in levels of noradrenaline in the hypothalamus (significant decreases of 29 and 40% in the mid- and high-dose groups) and of dopamine in the corpus striatum (significant decreases of 21, 26 and 35% in the low-, mid- and high-dose groups). There were dose-related, but not always statistically significant, decreases in all of the neurochemicals measured in the hypothalamus: noradrenaline, dopamine, vanillylmandelic acid (VMA), 3,4-dihydroxy-phenylacetic acid (dopac), homovanillic acid (HVA), serotonin (5-HT) and 5-hydroxyindoleacetic acid (5-HIAA). There were significant decreases in VMA in the midbrain, corpus striatum and cortex, 5-HT in the midbrain, corpus striatum and medulla oblongata, and dopac in the cerebellum in the high-dose group only. There were also significant decreases in 5-HT and 5-HIAA in the hypothalamus of the mid- and high-dose groups (Hsieh et al., 1992).

Continuous exposure of rats to 0.012, 0.12 and 5.3 mg phenol/m^3 for 61 days caused a shorter extensor muscle chronaxy and an increase in whole-blood cholinesterase activities in rats at concentrations above 0.012 mg phenol/m^3 (Mukhitov, 1964).

7.8.2 Myelotoxicity

Because phenol is an important metabolite of benzene, which is known to exert a toxic effect on bone marrow (including leukaemia) after metabolic activation, many studies have been performed to investigate the possible myelotoxic action of phenol.

In an *in vitro* assay for toxicity to primary murine haematopoietic cell cultures, phenol showed slight and variable activity at a concentration of 0.4 mM, but there was marked toxicity at 2 mM. In comparison, the phenol metabolites catechol and hydroquinone exhibited marked toxicity at 0.04 mM (Seidel et al., 1991).

Subcutaneous treatment with phenol (245 mg/kg body weight) significantly inhibited erythropoiesis in mice 48 h after treatment, as indicated by a ^{59}Fe uptake assay (Bolcsak & Nerland, 1983).

Intraperitoneal injection of 0-150 mg phenol/kg body weight to male B6C3F$_1$ mice, twice daily for 12 days, did not result in a suppression of bone marrow cellularity. However, simultaneous treatment of mice with 75 mg phenol/kg body weight and 25-75 mg hydroquinone/kg body weight (another benzene metabolite), produced a dose-related decrease in bone marrow cellularity, which was much more pronounced than after treatment with hydroquinone only. The observed effect closely resembled the myelotoxic effect of benzene (Eastmond et al., 1987b). Subsequent *in vitro* studies by these and other investigators confirmed that phenol (0.01-1 mM) stimulates further bioactivation of hydroquinone to myelotoxic compounds in bone-marrow cells (Eastmond et al., 1987b; Subrahmanyam et al., 1989).

No haematopoietic toxicity was found in rats after daily subcutaneous injections of 250-750 mg phenol/kg body weight for 1 week. Of the rats receiving 750 mg phenol/kg body weight, 50% died (Mitchell, 1972).

Six consecutive subcutaneous injections of 50 mg phenol/kg body weight to mice resulted in a slightly but significantly reduced number of granulopoietic stem cells and bone marrow cellularity in the tibia (Tunek et al., 1981).

7.8.3 *Immunotoxicology*

One immunological study has been reported. Female CD1 mice were exposed to 19 mg phenol/m^3 (5 ppm), either as a single 3-h exposure or as five daily 3-h exposures. Neither the susceptibility of the animals to experimentally induced streptococcus aerosol infection nor their pulmonary bactericidal activity was significantly affected (Aranyi et al., 1986).

Groups of five male CD-1 mice were supplied with drinking-water containing 0, 4.7, 19.5 or 95.2 mg phenol/litre for 4 weeks, at the end of which various haematological and immunological parameters were measured. The erythrocyte count was statistically significantly decreased, compared with control values, in all treated groups in a dose-related manner, but total and differential leucocyte counts were unaffected. Total spleen cellularity was decreased in a non-significant dose-related manner. The highest dose suppressed the stimulation of cultured splenic lymphocytes by the B-cell mitogen lipopolysaccharide, the T-cell mitogen phytohaemagglutinin, and the T- and B-cell mitogen pokeweed, but not by concanavatin. The mid and high doses suppressed the animals' antibody production in response to a T-cell-dependent antigen, i.e. sheep erythrocytes (Hsieh et al., 1992).

7.8.4 Biochemical effects

In a study on the biochemistry of intestinal mucosa, mice were provided with 0,5, 50 or 500 mg phenol/litre drinking-water (calculated by the authors to be 1, 10 and 100 mg phenol/kg body weight) for 5 days or 5 months at 1-day intervals. An increase in glucose-6-phosphatase, succinate dehydrogenase and cytochrome oxidase activities in the intestinal mucosa was observed in mice receiving $\geq$ 0.02 mg phenol/kg body weight for 5 days. A decrease in these activities was seen at 2 mg phenol/kg body weight. After 5 months administration of 0.02 and 0.2 mg phenol/kg body weight, the enzyme activity had returned to normal, but the highest dose group showed a decline (or even a lack) of activity in the cells of the intestinal mucosa (Olowska et al., 1980).

In another study of biochemical effects, mice were provided for 5 days (killed 24 h after the last application) and 35 days (killed 30 days after the last application) with aqueous solutions of 0.08, 0.8 or 8 mg phenol/litre. The authors of the study calculated the doses to be approximately 0.016, 0.16 and 1.6 mg phenol/kg body weight per day, respectively. Only the lowest dose of 0.08 mg phenol/litre evoked considerable changes in the localization of glucose-6-phosphatase, 5 days after treatment. Changes in alkaline phosphatase localization in the kidney were seen at 0.8 and 8 mg phenol/litre. Full recovery occurred after 30 days (Laszczynska et al., 1983).

Inhalation exposure of 50 male white rats to 0.4 mg phenol/m^3, 24 h/day, 7 days/week, for 3 months, resulted in some inhibition of oxidative phosphorylation in the lungs, liver and kidneys. An increase in the rate of glycolysis was also observed in the lungs and kidneys (Skvortsova & Vysochina, 1976).

8. EFFECTS ON HUMANS

8.1 General population exposure

8.1.1 Controlled studies

In the Kligman maximization test, phenol did not cause sensitization in 24 human volunteers (Kligman, 1966).

It was reported by Rea et al. (1987) that, in a group of 134 "chemically sensitive" patients where several volatile organic chemicals were detected in the blood, 107 (80%) reacted adversely after a challenge exposure to phenol alone (0.008 mg/m^3). The criteria used to identify "sensitive patients" and "adverse reactions" were not specified. The toxicological significance of this finding is not known.

Mukhitov (1964) reported that six 5-min inhalation exposures to phenol at 0.015 mg/m^3 produced an increased sensitivity to light in each of 3 dark-adapted subjects.

8.1.2 Case reports

Various reports have appeared on the adverse effects of phenol in individuals or groups of humans after intentional (e.g., therapeutic) as well as accidental short-term exposure to phenol.

8.1.2.1 Dermal exposure

The use of phenol as a disinfectant and antiseptic was introduced by Lister (1867). However, its use has been restricted by intoxications caused by these applications (Table 10).

Local effects after dermal phenol exposure consisted of erythema or painless blanching (Dreisbach, 1983), and, in more severe cases, corrosion (Schmidt & Maibach, 1981) and necrosis. The use of 5-10% phenol dressings for antiseptic purposes, for example, has led to many cases of necrosis of the skin and underlying tissues. When fingers and toes have been involved, amputation has sometimes been necessary. Due to their high toxicity, these dressings are no longer used (Cronin & Brauer, 1949; Deichmann, 1949; De Groote & Lambotte, 1960; Abraham, 1972).

Table 10. Human dermal toxicity of phenol

Concentration (%)	Medium	Contact duration	Circumstances	Most severe response	References
100	crystals	30 min	in glove	grangrene	Abraham (1972)
80-100	water	20 min	spill on hip, thigh, scrotum	death	Turtle & Dolan (1922)
80-100	water	2-4 days	closed dressings on open wounds	11 persons exposed: 1 death, 8 gas gangrene, 11 tissue necrosis	Lister (1867)
78	water	2-5 min	4-5 litres spilt on upper half of body	coma	Duverneuil & Ravier (1962)
43.5	waste water	1 min	spill on lower half of body	shock	Evans (1952)
5	ointment	7 days	closed dressing on cut	gangrene	Schussler & Stern (1911)
2	water	2.5 days	moist dressing over burns on 30% of body surface	death	Cronin & Brauer (1949)
2	water	11 h	closed bandage on infant umbilicus	death	Hinkel & Kintzel (1968)

Phenol chemical peel is a technique which has been used in superficial surgery of the skin for the last 30 years (Ersek, 1991). The phenolic mixture used classically is 3 ml of 50% phenol, 2 ml of water, 8 drops of soap and 8 drops of croton oil. This is applied to the skin to reduce pigmentation. Topical use of phenol as a chemical face-peel has been reported as being associated with cardiac dysrhythmias in "up to 30% of adults" (Morrison et al., 1991), but only a single case report has been published (Warner & Harper, 1985). This report concerned a 10-year-old boy who had a solution, consisting of 40% phenol and 0.8% croton oil in hexachlorophene soap and water, applied to a large nevus covering 1.9% of his body surface whilst under anaesthesia (60% nitrous oxide and 3% halothane) and receiving a total of 200 ml of lactated Ringer's solution intravenously. After 55 min of treatment, multifocal and coupled premature ventricular complexes were detected by ECG, but blood pressure remained stable and plasma sodium and potassium concentrations were normal. An intravenous infusion of 250 mg bretylium sulfate suppressed the dysrhythmia and the boy had an uneventful recovery.

Systemic intoxication can occur very rapidly after absorption of phenol through the skin (Table 10). Most significantly, cardiovascular shock (sometimes resulting in death) and severe metabolic acidosis occur. Truppman & Ellenby (1979) observed cardiac arrhythmias (supraventricular as well as ventricular) in 10 out of 42 patients within 10 min after the application of approximately 5% phenol on half of the face for cosmetic treatment. Hyperventilation, kidney damage and methaemoglobinaemia have also been observed in several cases of exposure to phenol.

Foxall et al. (1991) reported a case of acute renal failure following an industrial accident in which a man was partially submerged for a few seconds in a solution of 20% phenol in dichloromethane. He immediately showered, but was subsequently found in a state of collapse. His extremities were cold and he had 50% body burns. He developed nausea and vomiting after taking fluids. Anuria ensued, with a rise in plasma creatinine, but treatment with intravenous furosemide and haemodyalisis (daily for seven days, then with decreasing intervals for a further 18 days), allowed adequate urinary volumes to be produced. Respiratory distress required intensive care treatment. Marginal polyuria persisted one year after the accident.

8.1.2.2 *Oral exposure*

Cases of oral intoxication have occurred as a result of accidental and intentional ingestion. Local and systemic effects have been described in the literature, symptoms being similar to those following dermal exposure. Case reports have been published (Model, 1889; Stajduhar-Caric, 1968; Haddad et al., 1979).

Death occurred within 10 min of ingestion of 4.8 g phenol (Andersen, 1869). However, ingestion of 56.7 g of a phenol-saline mixture was reported to have occurred without complaints (Leider & Moser, 1961), and an individual survived the ingestion of 57 g phenol (88%) after intensive treatment. Symptoms in the latter study included severe gastrointestinal irritation, as well as the expected cardiovascular and respiratory effects (Bennett et al., 1950).

A severe accidental phenol spill in Wisconsin in 1974 contaminated ground water which was being used as drinking-water. Approximately one month later, several people living near the spill complained of health effects. Six months after the spill, medical histories were taken from 100 people who had consumed phenol-contaminated water (the authors estimated the daily exposure to be 10-240 mg phenol/person). In retrospect, a statistically significant increase was found in diarrhoea, mouth sores, dark urine and burning of the mouth, which had persisted for an average of 2 weeks. No significant abnormalities were found 6 months after initial exposure upon physical examination or laboratory analysis. Urinary phenol levels were not elevated (Delfino & Dube, 1976; Baker et al., 1978).

A river in North Wales, United Kingdom, used for the preparation of drinking-water, was accidentally contaminated with phenol (Jarvis et al., 1985). When the water was chlorinated, various chlorophenols appeared to have been formed. A retrospective postal survey of 344 households that received the contaminated tap water and 250 control households was carried out. Significantly more gastrointestinal illnesses, as well as other symptoms, were claimed in the contaminated areas than in the unexposed areas. Phenol concentrations in drinking-water were conservatively estimated to have been 4.7-10.3 μg/litre for some days (Jarvis et al., 1985).

8.1.2.3 *Inhalation exposure*

Very few cases of adverse effects after short-term phenol vapour exposure have been reported.

Hospital outbreaks of severe idiopathic neonatal unconjugated hyperbilirubinaemia have been associated with the phenol-containing disinfectant used for cleaning the nursery equipment, floors and walls. When the disinfectant was no longer used, the epidemic subsided (Daum et al., 1976; Wysowski et al., 1978; Doan et al., 1979).

Studies of occupational inhalation exposure are described in section 8.2.

8.1.2.4 *Exposure by injection*

Phenol has been used as a neuron blocking agent in patients suffering from spasm following, for example, spinal cord damage or cerebrovascular stroke (Wood, 1978, review; Nathan, 1959; Cooper et al., 1965; Khalili & Betts, 1967; Gibson, 1987). It has also been used to relieve chronic pain (Wood, 1978; Benzon, 1979; Smith, 1984). Treatment involved administering the phenol by intravenous injection or perfusion, or by direct injection into the spinal cord. Reported side-effects of phenol therapy were convulsions, transient paraesthesia, leg weakness, urinary and fecal incontinence, one case of a severe arterial block in the upper arm requiring amputation, and one case of acute bronchospasm (Wood, 1978, review; Benzon, 1979; Gibson, 1987; Atkinson & Skupak, 1989). In addition, there have been reports of phenol-induced cardiac dysrhythmia in adults (Forrest & Ramage, 1987) and in children, in whom an incidence of 19% was reported (Morrison et al., 1991).

8.2 Occupational exposure

Poisoning due to chronic inhalation of phenol was known 100 years ago, primarily as a disorder in physicians and their helpers, under the term "carbol marasmus" (Lister, 1867). A classical case of phenol marasmus was described in a worker employed for $13\frac{1}{2}$ years in a laboratory boiling phenol solutions. Symptoms were anorexia, weight loss, headache, vertigo, salivation and dark urine (Merliss, 1972).

A few studies are available concerning occupational exposure of workers in bakelite factories. Workers were exposed to phenol, and simultaneous exposure to formaldehyde occurred. Elevated phenol urine levels, unspecified complaints, and chronic airway obstruction were observed (Schoenberg & Mitchell, 1975; Knapik et al., 1980).

Twenty-nine cases of poisonings among workers, who, during a 3-year period, quenched coke with a waste-water solution containing 0.3-0.8 g phenol/litre, were attributed to phenol intoxication. Phenol vapour concentrations in the air ranged from 0.5 to 12.2 mg/m^3. The number of workers and the symptoms of intoxication were not specified. The author did not consider the potential of dermal absorption (Petrov, 1960).

A case-control study was carried out on 57 cases among 3805 workers from the Finnish wood industry (particle board, plywood, sawmill or formaldehyde glue) suffering from respiratory cancer. The inhalatory exposure level to phenol and frequency of multiple exposure to pesticides were found to be significantly higher for cancer cases, but the exposure to wood dust was not significantly different between cases and controls (Kauppinen et al., 1986). The number of cases in this study was small, and confounding exposures were inadequately controlled.

In a case-control study of 6678 rubber workers, employed in areas where phenol was used, exposure to phenol was not associated with increased risks of cancer of the respiratory tract, stomach or prostate or of lymphosarcoma or lymphatic leukaemia (Wilcosky et al., 1984).

A mortality study was conducted among 14 861 white male workers engaged in the production or use of phenol and formaldehyde in five companies within the USA. The follow-up comprised more than 360 000 man-years. Mortality rates from all causes combined were similar to those in the general population of the USA. Excesses of cancer of the oesophagus, cancer of the kidney and Hodgkin's disease were observed among the workers exposed to phenol, but these did not show any exposure-response relationship and were not statistically significant. Reduced mortality ratios were observed for cancer of the buccal cavity and pharynx, cancer of the stomach, cancer of the brain, arteriosclerotic heart disease, emphysema, disease of the digestive system and cirrhosis of the liver, although these reductions were not statistically significant. For arteriosclerotic heart disease, emphysema and cirrhosis of the liver, there were inverse relationships between mortality rates and duration of phenol exposure and cumulative phenol exposure levels (Dosemeci et al., 1991).

A cardiovascular disease (CVD) mortality study was conducted among 1282 white male production workers in a large rubber - and tyre-manufacturing plant. Exposure estimates for 25 solvents were available (concentrations were not measured). The CVD mortality during 15-year follow-up period was analyzed in exposed and not exposed workers. The known association between CS_2 exposures and ischemic heart diseases (IHD) was confirmed, and two other solvents, ethanol and phenol, were also found to be predictors of IHD. Phenol showed the strongest association with CVD mortality. However, some confounders (cigarette smoking, hypertension and high serum cholesterol) were not controlled and unrecognized chemical atherogens could also, according to the authors, influence the results (Wilcosky & Tyroler, 1983).

8.3 Organoleptic data

The odour threshold for phenol has been reported to range from 0.021 to 20 mg/m³ (Van Gemert & Nettenbreijer, 1977; Van Gemert, 1984).

The geometric mean of 16 air odour thresholds and 6 water odour thresholds for phenol was reported by Amoore & Hautala (1983) to be 0.16 mg/m3 (0.040 ppm, with a standard error of 0.026 ppm). In this calculation, the original literature was reviewed and values which diverged more than 100-fold from the nearest of two or more other thresholds were eliminated. Both detection and recognition values were included. The water detection threshold for phenol, based upon multiplying the calculated air odour threshold by the water-air distribution ratio, was reported by the same authors to be 7.9 mg/litre.

A taste threshold value of 0.3 mg/litre in water has been reported (US EPA, 1992).

9. EFFECTS ON OTHER ORGANISMS IN THE LABORATORY AND FIELD

The toxicity of phenol has been studied in microorganisms (e.g., bacteria, fungi, algae and protozoa) and numerous aquatic invertebrates and vertebrates (Buikema et al., 1979). Because of this vast amount of data, a selection has been made, based on the reliability of the data and the relevance of the test organisms. Details of acute and long-term aquatic toxicity studies, considered to be adequately performed and reported, are included in Tables 11 and 12. Less adequate studies are reported in the text only.

9.1 Microorganisms

Reliable phenol toxicity data for microorganisms are given in Table 11.

In microorganisms, growth inhibition is usually observed after phenol exposure. In studies on single bacterial species, the EC_{50} values (EC_{50} = calculated concentration affecting 50% of test population) found for growth inhibition varied from 244 mg phenol/litre in a newly developed, 6-h test with *Pseudomonas putida* (Slabbert, 1986) to 1600 mg phenol/litre after 18 h of exposure in a more conventional test with *Aeromonas hydrophila* (Dutka & Kwann, 1981). Bringmann & Kühn (1977) reported a toxicity threshold of 64 mg/litre after 16 h. EC_{50} values for reduced photoluminescence in *Photobacterium phosphoreum* of 28-34 mg phenol/litre (Dutka et al., 1983) and 40 mg phenol/litre (Curtis et al., 1982) have been reported. In activated sludge, the EC_{50} for a reduced oxygen uptake was reported to be 520-1500 mg phenol/litre, whereas a lower value was found for substrate consumption inhibition (104 mg phenol/litre) (Miksch & Schürmann, 1988; Volskay & Grady, 1988). The lowest reported concentration affecting activated sludge was 10 mg phenol/litre; 1 mg phenol/litre had no effect (Baird et al., 1974).

Reported toxicity thresholds for protozoa were of the same order of magnitude as for bacteria: 33-144 mg phenol/litre (Bringmann & Kühn, 1959, 1980; Dive & LeClerq, 1977; Bringmann et al., 1980). For algae, values were somewhat lower, but were observed after a longer exposure period: 6 mg phenol/litre for cyanobacteria (blue-green algae) and 8 mg phenol/litre for green algae, after 7-8 days of exposure (Bringmann & Kühn, 1978, 1980). The IC_{50} values (concentration

Table 11. Acute aquatic toxicity of phenol

Organism	Temperature (°C)	pH	Dissolved oxygen (mg/litre)	Hardness (mg CaCO$_3$/litre)	Method[a]	Test duration	Parameter[b]	Concentration (mg/litre)	Reference
Freshwater Organisms									
Bacteria									
Photobacterium phosphoreum	15	6.5-6.7			S	5, 10, 15 min	EC$_{50}$[c]	28, 32, 34	Dutka et al. (1983)
Pseudomonas putida	25	7.0		80.1	S	16 h	TT[c]	64	Bringmann & Kuhn (1977)
	27	7.2			S	6 h	EC$_{50}$	244	Slabbert (1986)
Cyanobacteria									
Microcystis aeruginosa	27	7.0		72.3	S	8 days	TT	6	Bringmann & Kuhn (1978)
Green algae									
Scenedesmus quadricauda	27	7.0		72.3	S	7 days	TT	8	Bringmann & Kuhn (1980)
Protozoa									
Chilomonas paramaecium	20	6.9			S	48 h	TT	65	Bringmann et al. (1980)

Table 11 (contd).

Organism	Temperature (°C)	pH	Dissolved oxygen (mg/litre)	Hardness (mg CaCO3/litre)	Method[a]	Test duration	Parameter[b]	Concentration (mg/litre)	Reference
Colpidium campylum	20				S	43 h	TT[d]	100	Dive & LeClerq (1977)
Entosiphon sulcatum	25	6.9		80.1	S	72 h	TT	33	Bringmann & Kuhn (1980)
Microregma heterostoma	27	7.5-7.8		213.6	S	28 h	TT	30[e]	Bringmann & Kuhn (1959)
Uronema parduczi	25	7.3			S	20 h	TT	144	Bringmann & Kuhn (1980)
Crustacea									
Asellus aquaticus	11 ± 1	7.5-8.1		99.5 ± 7.7	S	24, 48, 96 h	LC_{50}	230, 200, 180[f]	Green et al. (1985)
Cypris subglobosa	20.4	7.9	8.4	204	S	12, 24, 48, 72, 96 h	LC_{50}	173, 167, 137, 122, 72	Rao et al. (1983)
Daphnia magna	19.8-20.9	7.7-8.3		157 ± 4	S	48 h	LC_{50}	13	Gersich et al. (1986)

Table 11 (contd).

Species	Temp (°C)	pH		Hardness	S/R	Time	Endpoint	Values	Reference
	19 ± 1	8.2 ± 0.3		199.4	S	48 h	LC_{50}	100[f]	Hermens (1984)
	19 ± 1	8.2 ± 0.3		199.4	S	48 h	EC_{50}[g]	9[f]	Hermens (1984)
	17.2 ± 0.5	7.4 ± 0.2	8.7 ± 1.1	44.7	S	24 h	LC_{50}[h]	13	Holcombe et al. (1987)
	22 ± 1	6.8-7.8	7.6 ± 0.2	146 ± 15	S	48 h	EC_{50}[g]	7	Keen & Baillod (1985)
					S	48 h	LC_{50}	8	Lewis (1983)
Gammarus pulex	11 ± 1	7.5-8.1		99.5 ± 7.7	S	24, 48, 96 h	LC_{50}	106, 85, 69[f]	Green et al. (1985)
	7 ± 1	8.3	10.9	250	R	24, 48, 72, 96 h	LC_{50}	100, 89, 67, 51	Stephenson (1983)
Ceriodaphnia dubia	25 ± 1	8.18 ± 0.04		57.1 ± 4.1	S	48 h	LC_{50}	3.1	Oris et al. (1991)
Mollusca *Indoplanorbis exustus*					S	12, 24, 48, 72, 96 h	LC_{50}	265, 215, 200, 156, 126	Agrawal (1987)
Lymnaea acuminata	20 ± 2	7.9 ± 0.2	5.5 ± 1.5	190-223	R	12, 24, 48, 72, 96 h	LC_{50}	270, 219, 205, 158, 129	Gupta & Rao (1982)
Worms *Limnodrilus hoffmeisteri*	11 ± 1	7.5-8.1		99.5 ± 7.7	S	24, 48, 96 h	LC_{50}	960, 870, 780[f]	Green et al. (1985)

Table 11 (contd).

Organism	Temperature (°C)	pH	Dissolved oxygen (mg/litre)	Hardness (mg CaCO3/litre)	Method[a]	Test duration	Parameter[b]	Concentration (mg/litre)	Reference
Polycelis felina	18	7-8.5	[j]	300-500	S	96 h	LC$_{50}$	64[f]	Erben et al. (1983)
Polycelis tenuis	11 ± 1	7.5-8.1		99.5 ± 7.7	S	24, 48, 96 h	LC$_{50}$	230, 200, 88[f]	Green et al. (1985)
Fish									
Brachydanio rerio	25 ± 0.5	8.0-8.3	[j]	350-375	CF	48, 96 h	LC$_{50}$	31, 29[f]	Fogels & Sprague (1977)
	24 ± 1		> 6	64	R (12 h)	6, 12, 24, 48, 72, 96 h	LC$_{50}$	35, 31, 28, 26, 25, 25[f]	Razani et al. (1986a)
Campostoma anomalum	23		[j]		R (24 h)	48 h	LC$_{50}$	18[f]	Chagnon & Hlohowskyj (1989)
Catostomus commersoni	17.2 ± 0.5	7.4 ± 0.2	8.7 ± 1.1	44.7	S	96 h	LC$_{50}$[h]	11	Holcombe et al. (1987)
Jordanella floridae	25 ± 0.5	8.0-8.3	[j]	350-375	CF	48, 96 h	LC$_{50}$	36, 36[f]	Fogels & Sprague (1977)
Lebistes reticulatus	28-31.8	7.8-8.2	5.7-7.2	218-239	R (24 h)	12, 24, 48, 72, 96 h	LC$_{50}$	103, 83, 64, 50, 48	Gupta et al. (1982a)

Table 11 (contd).

Lepomis macrochirus	17.2 ± 0.5	7.4 ± 0.2	8.7 ± 1.1	44.7	S	96 h	LC_{50}[h]	17	Holcombe et al. (1987)
Leuciscus idus melanotus	20				S	48 h	LC_{50}	14, 25	Jünke & Lüdemann (1978)
Notopterus notopterus	23-26.5	6.8-7.6	5.9-7.8	60-70	S	24, 48, 72, 96 h	LC_{50}	14, 14, 13, 13	Verma et al. (1980)
Pimephales promelas	17.2 ± 0.5	7.4 ± 0.2	8.7 ± 1.1	44.7	S	96 h	LC_{50}[h]	25	Holcombe et al. (1987)
	25 ± 2		6.2-8.2	43-49	CF	96, 192 h	LC_{50}	29, 23[f]	Phipps et al. (1981)
Rasbora heteromorpha	20	7.2		250	S	24, 48 h	LC_{50}	8, 7	Alabaster (1969)
Rutilus rutilus	10.3 ± 0.3	7.8 ± 0.02		257-260	S	48 h	LC_{50}	10[f]	Solbé et al. (1985)
Salmo gairdneri	15 ± 0.5	8.0-8.3	–	350-375	CF	48, 96 h	LC_{50}	12 12[f]	Fogels & Sprague (1977)
	17.2 ± 0.5	7.4 ± 0.2	8.7 ± 1.1	44.7	S	96 h	LC_{50}[i]	11	Holcombe et al. (1987)
Insects									
Baetis rhodani	11 ± 1	7.5-8.1		99.5 ± 7.7	S	24, 48, 96 h	LC_{50}	19, 19, 16[f]	Green et al. (1985)

Table 11 (contd).

Organism	Temperature (°C)	pH	Dissolved oxygen (mg/litre)	Hardness (mg CaCO3/litre)	Method[a]	Test duration	Parameter[b]	Concentration (mg/litre)	Reference
Chironomus riparius	11 ± 1	7.5-8.1		99.5 ± 7.7	S	24, 48, 96 h	LC_{50}	1050, 500, 240[f]	Green et al. (1985)
Hydropsyche angustipennis	11 ± 1	7.5-8.1		99.5 ± 7.7	S	24, 48, 96 h	LC_{50}	940, 720, 260[f]	Green et al. (1985)
Marine Organisms									
Crustacea									
Artemia salina					S	24, 48 h	LC_{50}[j]	157, 56	Price et al. (1974)
Canthocamptus	synthetic medium according to Cairns				S	48 h	LC_{50}[j]	9	Rao & Nath (1983)
Gammarus duebeni	5 ± 0.6	7.7 ± 0.1	8.4 ± 0.3	0.6%[k]	CF	96 h	LC_{50}	183[f]	Oksama & Kristofferson (1979)
	16 ± 0.6	7.7 ± 0.1	8.4 ± 0.3	0.6%[k]	CF	96 h	LC_{50}	89[f]	
Mesidotea entomon	5 ± 0.6	7.7 ± 0.1	8.4 ± 0.3	0.6%[k]	CF	96 h	LC_{50}	176[f]	Oksama & Kristofferson (1979)
	10 ± 0.6	7.7 ± 0.1	8.4 ± 0.3	0.6%[k]	CF	96 h	LC_{50}	186[f]	
Panopeus herbstii	25			25 ppt[k]	R 24 h	96 h	LC_{50}	53	Key & Scott (1986)

Table 11 (contd).

								Reference
Mollusca								
Crassostrea virginica	24 ± 1		sea[k]	R	48 h (eggs)	LC$_{50}$	58	Davis & Hidu (1969)
Mercenaria mercenaria	24 ± 1		sea[k]	R	48 h (eggs)-12 days (larvae)	LC$_{50}$[j]	53-55	Davis & Hidu (1969)
Worms								
Ophryotrocha diadema	21		sea[k]	S	48 h	LC$_{50}$	100-330	Parker (1984)
Fish								
Phoxinus phoxinus	5 ± 0.7	7.7 ± 0.1 8.4 ± 0.3	0.6%[k]	CF	96 h	LC$_{50}$	10	Oksama & Kristofferson (1979)

[a] S = static test; CF = continuous flow test; R = renewal test (semi-static test)
[b] EC$_{50}$ = median effect concentration = calculated concentration causing effect in 50% of population; LC$_{50}$ = median lethal concentration
      TT = toxicity threshold, i.e. concentration affecting growth in ≥ 3% of population
[c] effect: reduction in photoluminescence
[d] minimal concentration affecting growth
[e] river water
[f] concentration of test compound analysed during assay
[g] immobility
[h] simultaneous testing of 8 species
[i] no data; aerated
[j] reported was TL$_m$ or median toxicity limit
[k] salinity; sea = sea water

Table 12. Long-term aquatic toxicity of phenol

Organism	Temperature (°C)	pH	Dissolved oxygen (mg/litre)	Hardness (mg/litre CaCO₃)	Method[a]	Test duration	Parameter[b]	Concentration (mg/litre)	Reference
Freshwater Organisms									
Crustacea									
Daphnia magna	19 ± 1	8.2			199	R (48 h)	16 days	NOEC (growth)	0.16	De Neer et al. (1988)
	19 ± 1	8.2 ± 0.3			199	R	21 days	NOLC NOEC (repro) NOEC (growth)	10[c] 3.2[c] 3.2[c]	Hermens (1984)
Ceriodaphnia dubia	25 ± 1	8.18 ± 0.04		57.1 ± 4.1	R (48 h)	4 days 7 days	NOEC (repro) NOEC (survival)[d]	4[c] 4[c]	Oris et al. (1991)
Fish									
Brachydanyo rerio	24 ± 1	6.1-6.5	6.4-8.5	57-61	R (24 h)	3 months (adults)	NOLC	4.9	Razani et al. (1986b)
					R (48 h)	2 months (test started with eggs from adults exposed to same concentrations; all eggs hatched after 2-3 days)	NOEC (spawning) NOEC (growth) NOLC	< 2.2 2.2 < 2.2	

Table 12 (contd).

Species									
Carassius auratus	19-24	7.78	6.2-9.0	197.5	CF	8 days after hatching (test was started with eggs (1-2 h after spawning) which hatched completely within 3.5 days)	LC_1 (at hatch)	0.002^c	Birge et al. (1979)
Pimephales promelas	25 ± 1	8.0	5.3	725.3	CF	30 days after hatching	NOEC (hatching)	83.2^c	De Graeve et al. (1980)
							NOEC (growth)	0.75^c	
Pimephales promelas	25 ± 2	7.2-7.9	5.3	725.3	CF	38 days after hatching (test was started with eggs within 24 h after spawning)	NOLC	$6.1^{c,e}$	Holcombe et al. (1982)
			7.7	46.0	CF		NOLC	$3.57^{c,f}$	
							NOEC (growth)	1.83^c	
							NOEC (hatching)	$3.57^{c,f}$	
Salmo gairdneri	12-14	7.78	6.2-9.0	197.5	CF	8 days after hatching (test was started with eggs (20 min after fertilization) which hatched completely within 22 days)	LC_1 (at hatch)	0.0002^c	Birge et al. (1979)

Table 12 (contd).

Organism	Temperature (°C)	pH	Dissolved oxygen (mg/litre)	Hardness (mg/litre CaCO$_3$)	Method[a]	Test duration	Parameter[b]	Concentration (mg/litre)	Reference
	13.3-14.2	7.4-8.1	8.6-10.2	100	CF	4 days after hatching (test was started with eggs (20 min after fertilization) which hatched within 23 days)	NOLC NOEC (hatching)	0.009[c,e] 0.009[c,f]	Black et al. (1983)
		7.8	5.7	579.9	CF	58 days after hatching (test was started with eyed eggs which hatched completely within 48 h)	NOEC (growth)	0.1[c,g]	De Graeve et al. (1980)

[a] S = static test; CF = continuous flow test; R = renewal test (semi-static)
[b] NOEC = no-observed-effect concentration = highest tested concentration without observed effect; NOLC = no-observed-lethal concentration.
     An LC$_1$ is a calculated value which is, to some extent, comparable to the observed NOLC value used in other studies.
[c] phenol concentration analysed during test
[d] survival was a more sensitive end-point than reproduction
[e] calculated from results by Task Group
[f] highest concentration tested
[g] extrapolated by authors

92

causing 50% growth inhibition) reported for various fungi by Kwasniewska & Kaiser (1983) were of the same order of magnitude as the above EC_{50} values for bacterial growth inhibition: 460-1000 mg phenol/litre. These values are also within the range of concentrations observed by Babich & Stotzky (1985) to cause initial or complete growth inhibition in various fungi (100-1000 mg phenol/litre and 750->1000 mg phenol/litre, respectively).

An increase in salinity (0-30%) increased the toxicity of phenol to fungi (Babich & Stotzky, 1985).

9.2 Aquatic organisms

9.2.1 Freshwater organisms

9.2.1.1 Short-term studies

The most important sublethal acute effects observed in freshwater species after phenol exposure were a reduced heart rate and damage to the epithelium of gills (with loss of function), liver, kidneys, intestines and blood vessels. One study reported the occurrence of severe seizures, mediated by the central nervous system, in *Salmo gairdneri* after exposure to sublethal phenol concentrations (Bradbury et al., 1989). In invertebrates, growth inhibition was usually observed. Some EC_{50} values for the latter organisms are given in Table 11.

Most toxicity studies concentrated on lethal effects. Death was usually preceded by immobility, loss of equilibrium, paralysis and respiratory distress (Razani et al., 1986a; Tonapi & Varghese, 1987; Green et al., 1988; Chagnon & Hlohowskyj, 1989). Toxicity testing, where the same species was used by different research workers in different waters, resulted in LC_{50} values (LC_{50} : calculated concentration causing death in 50% of test group) that varied widely, as can be seen from Table 11.

Environmental factors may affect the toxicity of phenol (Brown et al., 1967; Miller & Ogilvie, 1975; Ruessink & Smith, 1975; Cairns et al., 1976; Reynolds et al., 1978; Birge et al., 1979; Dalela et al., 1980; Gluth & Hanke, 1983; Gupta et al., 1983a,b; Stephenson, 1983). Hardness and pH, however, do not appear to have a large impact on phenol toxicity. The toxicity for various fungi and fish species, for example, did not change significantly over the pH range of 5-8; toxicity for fungi and some fish species was not influenced at all by hardness, whereas phenol was slightly more toxic in soft than in hard waters for the carp (Herbert, 1962; Pickering & Henderson, 1966; EIFAC, 1972; Babich & Stotzky, 1985). The effect of temperature appeared to be variable (Cairns et al., 1978; Babich & Stotzky, 1985). Since temperature influences both the uptake and the detoxification (conjugation) of phenol (Green et al., 1988), phenol toxicity could be enhanced, as well as diminished, by increasing temperature, depending on which parameter was influenced most.

Several biological factors also influence the response of the biota to phenol, e.g., strain type, nutritional status, size, embryonal or developmental stage, crowding and physiological adaptation (Dowden & Benett, 1965; Alexander & Clarke, 1978; Birge et al., 1979; Flerov, 1979; De Graeve et al., 1980; Kordylewska, 1980; Gupta et al., 1982b; Mayes et al., 1982; Black et al., 1983; Lewis, 1983).

Comparison of 48-h LC_{50} values from Table 11 shows that, in general, fish are the most sensitive freshwater species with respect to phenol toxicity. The 48-h LC_{50} values for some selected fish species ranged from 7 to 64 mg/litre. For crustaceans, this range was 3.1-200, and for molluscs it was 200-205 mg/litre; for insects, it was 19-720 mg/litre and for worms 200-870 mg/litre. Upon simultaneous testing of eight species at concentrations up to 51 mg phenol/litre, no toxicity was observed for larvae of the amphibian *Xenopus laevis*, for the snail *Aplexa hypnorum* or the insect *Tanytarsus dissimilis*. Where toxicity was observed, LC_{50} values were included in Table 11 (Holcombe et al., 1987). The data presented in Table 11 are in good agreement with the order of increasing tolerance to phenol proposed by Alekseyev & Antipin (1976): fish-crustaceans-tolerant insects-worms-molluscs-highly tolerant insects.

9.2.1.2 Long-term studies

Most long-term studies with freshwater species have concerned growth, reproduction and/or mortality; these studies are discussed below. Studies considered to be adequately performed and reported are included in Table 12.

A few long-term studies with freshwater fish have been designed to detect sublethal effects of phenol exposure. Increased proteolysis as a result of stress, mild kidney damage, and an inhibitory effect on the development and maturation of the ovary, secondary to a liver dysfunction, were some of the effects reported (Dangé, 1986; Gupta & Dalela, 1987; Kumar & Mukerjee, 1988).

In a life-cycle test using *Daphnia magna*, the maximum acceptable tolerance concentration (MATC) proved to be 1.5-6.3 mg phenol/litre (US EPA, 1980). These results are comparable with the no-observed-effect concentration (NOEC) values for growth and reproduction (both 3.2 mg phenol/litre) found by Hermens (1984) and Oris et al. (1991). De Neer et al. (1988), however, found a considerably lower NOEC value of 0.16 mg phenol/litre for growth of *Daphnia magna* under similar experimental conditions.

Exposure of adult *Brachydanio rerio* to 2.2, 4.9 or 24 mg phenol/litre for 3 months resulted in 67% mortality at the highest concentration; the no-observed-lethal concentration (NOLC) was 4.9 mg phenol/litre. At 24 mg phenol/litre, only immature oocytes were found in surviving fish; at the two lower concentrations both immature and mature oocytes were observed, whereas spawning was delayed. In subsequent embryo-larval tests, starting with the eggs of exposed adults, mortality appeared to be maximal during embryonic development and the initial larval stage. All larvae died within 12 days at 4.9 mg phenol/litre. At 2.2 mg phenol/litre, larval mortality was still slightly increased, but surviving animals showed normal growth and development (Razani et al., 1986b).

In two embryo-larval bioassays on *Pimephales promelas*, growth proved to be the most sensitive criterion: the NOEC values were 0.75 and 1.83 mg phenol/litre (De Graeve et al., 1980; Holcombe et al., 1982).

The results of the embryo-larval test on *Salmo gairdneri* given by Birge et al. (1979) (LC_1:0.2 μg/litre) and Black et al. (1983) (NOEC:9 μg/litre) are much lower than those obtained by De Graeve et al. (1980) ($NOEC_{growth}$: 0.1 mg/litre), probably because the latter test was started with eyed eggs, whereas the former two tests were started with just-fertilized eggs.

In addition, Birge et al. (1979) studied the influence of phenol in an embryo-larval test on *Carassius auratus* and found a LC_1 of 2.0 μg phenol/litre.

Dumpert (1987) reported a NOLC value of 10 mg phenol/litre for larvae of the amphibian *Xenopus laevis*; larval mortality was 100% within 3 weeks at 50 mg phenol/litre. Larval growth was slightly, but not significantly, retarded at 5 and 10 mg phenol/litre. Hatching was normal at all tested concentrations. However, the results may not be reliable because test solutions were renewed only once a week, whereas aeration may also have contributed to undetected loss of phenol. In other embryo-larval bioassays on five amphibian species, *Rana ripiens* and *Rana catesbeiana* were the least tolerant. These species exhibited about equal sensitivity to phenol (LC_1: 1.0 and 1.1 mg phenol/litre, respectively; LC_{10}: 5.2 and 8.5 mg phenol/litre, respectively (Birge et al., 1980).

9.2.2 Marine organisms

9.2.2.1 Short-term studies

In acute toxicity studies on some marine organisms (crustaceans, worms, snails and fish), the LC_{50} values ranged from 8.8-330 mg phenol/litre (see Table 11). In general, the sensitivities of marine and freshwater organisms for phenol were similar.

At a sublethal phenol concentration, activities of some enzymes appeared to be decreased in the brain, liver and muscle tissue of *Sarotherodon mossambicus*; this effect was independent of salinity (Ravichandran & Anantharaj, 1984).

9.2.2.2 Long-term studies

No adequate data are available on long-term toxicity to marine organisms.

9.2.3 Accumulation

The bioconcentration factor of phenol may be calculated, using a log P_{ow} value of 1.46 (pH not stated, see Table 1) and the formula log BCF = 0.79 log P_{ow} -0.40 (Veith & Kosian, 1983). This yields a bioconcentration factor of 5.7, which is very low and does not indicate any potential for bioaccumulation.

The experimental bioconcentration factors, reported by Hardy et al. (1985) for algae, by Erben (1983) for flatworms, by Erben

10. EVALUATION OF HUMAN HEALTH RISKS AND EFFECTS ON THE ENVIRONMENT

10.1 Evaluation of human health risks

10.1.1 Exposure

The main way in which the general population can be exposed, on a long-term basis, to phenol present in ambient air is as a result of industrial emissions and various combustion processes. Other inhalation sources include the decomposition of organic materials, liquid manure, and the atmospheric degradation of benzene. Inhalation and dermal exposure may arise from contact with contaminated water or consumer products containing phenol. Indirect exposure of man through the food chain is not likely to add significantly to long-term inhalation exposure, in view of the short life-time of the compound in the environment (section 10.2.1.). Individuals may ingest phenol via drinking-water from contaminated surface water or ground water. Repeated oral exposure may arise from the consumption of smoked food items. Endogenous production of phenol may be influenced by the diet and exposure to certain drugs and other xenobiotics.

The exposure data available are inadequate to determine the degree of exposure of the general population or of specific groups at risk, including workers.

An upper-limit estimate of the daily intake can be made for long-term exposure of the general population. In this hypothetical case, it is assumed that an individual will be maximally exposed to phenol through continuous inhalation of air from a heavily industrialized area, with frequent consumption of smoked food items with a high phenol content, and of drinking-water containing phenol up to the taste threshold. The estimate is summarized in the table below:

Source	Quantity of source	Phenol concentration	Phenol intake
Air	20 m^3/day	200 $\mu g/m^3$	4 mg/day
Smoked food items	200 g/week	70 mg/kg	2 mg/day
Drinking-water	2 litres/day	300 μg/litre	0.6 mg/day

Assuming an average body weight of 70 kg, the total daily intake of this maximally exposed individual will be 0.1 mg/kg body weight per day. The daily intake by the general population can be expected to be much less than this figure.

10.1.2 Toxicity

Phenol has moderately acute toxicity for animals. The oral LD_{50} for various animal species range from 300 to 600 mg phenol/kg body weight, and the LC_{50} for rats by inhalation is more than 900 mg phenol/m^3.

In humans, the lowest acutely lethal oral dose was reported to be 4.8 g, which is approximately 70 mg/kg body weight. Local, as well as systemic, effects have been reported in humans, consisting of irritation, necrosis, cardiovascular effects, metabolic acidosis, neurological effects and methaemoglobinaemia. Several fatal cases have been reported after oral or dermal intoxication. No documented cases of death by inhalation of phenol have been found.

Solutions of phenol are corrosive to the skin and eyes. Phenol vapours can irritate the respiratory tract. Phenol is not a skin sensitiser in guinea-pigs or humans.

The most important effects reported in short-term animal studies were neurotoxicity, liver and kidney damage, respiratory effects and growth retardation. Toxic effects in rat kidney have been reported to occur at oral dose levels of 40 mg/kg per day or more. Liver toxicity was evident in rats administered at least 100 mg/kg per day. In a limited 14-day study on rats, an oral NOAEL of 12 mg/kg per day was reported based on kidney effects. In this experiment, miosis (an iris response to light) was inhibited at 4 mg/kg per day (the lowest dose tested). However, the health significance of this finding is not clear. Some biochemical changes have been reported to occur in the intestinal mucosa and kidneys of mice at dose levels below 1 mg/kg per day. The toxicological value of these insufficiently reported biochemical observations is not known.

There have been no long-term general toxicity studies in animals or adequate epidemiological studies.

No adequate studies on the reproductive toxicity of phenol have been reported. Phenol has been identified as a developmental toxicant in studies with rats and mice. In two multiple-dose rat studies, NOAELs of 40 mg/kg per day (the LOAEL was 53 mg/kg per day) and 60 mg/kg per day (the LOAEL was 120 mg/kg per day) have been reported. For the mouse, the NOAEL was 140 mg/kg per day (the LOAEL was 280 mg/kg per day).

There is some evidence that phenol is genotoxic to mammalian cells *in vitro*. Based on the induction of bone marrow micronuclei in several studies with mice, phenol may have genotoxic potential.

Oral (drinking-water) animal carcinogenicity bioassays did not give evidence of a carcinogenic potential of phenol. No animal inhalation or adequate dermal carcinogenicity studies are available. Two-stage carcinogenicity studies with mice showed that phenol applied to the skin does have tumour-promoting activity. Adequate human data on carcinogenicity are not available.

10.1.3 Evaluation

Accidental high exposure to phenol may cause severe local effects, systemic intoxications and even death. The available data do not suggest a strong potential for cumulative health effects from chronic exposure.

The lowest NOAELs identified are for kidney and developmental effects, and in rats are in the range of 12-40 mg/kg body weight per day. The Task Group decided to derive a tolerable daily intake (TDI), taking into consideration this range. An uncertainty factor of 200 (including factors of 10 for interspecies variation, 10 for intraspecies variation, and 2 to account for the limited data base on the toxicity of phenol in animals) was considered appropriate. A range of 60-200 μg/kg per day was recommended as the upper limit of the TDI by the Task Group. As the Task Group's upper-limit estimate of human daily intake is 100 μg/kg body weight per day (section 10.1.1), it can be concluded that the average general population exposure to phenol from all sources will be well below this range.

There remain, however, two reasons for concern. The available data suggest that phenol may be genotoxic, and there is insufficient data to discount the possibility that phenol is carcinogenic. For these reasons, it is particularly important that this evaluation of phenol be kept under periodic review.

10.2 Evaluation of effects on the environment

10.2.1 Environmental levels

Once released into the environment, intercompartmental transport of phenol may occur by wet deposition from air to sea water and surface water and soil, and, as the compound can be expected to be highly mobile in soil, by leaching through soil. Evaporation will be slow from water and can only be expected following contamination of relatively dry soil.

Phenol, however, is generally not likely to persist in either air, sea or surface water, soil or sewage. It readily reacts photochemically and is rapidly biodegraded aerobically, mainly to carbon dioxide. Anaerobic degradation to carbon dioxide or methane also occur. Half-lives will be in the range of several hours for photodegradation and in the range of hours to days for aerobic biodegradation. Anaerobic biodegradation also occurs, albeit at a slower rate. Low removal rates of phenol in ground water and soil may occur, e.g., following spills, with subsequent inhibition of the microbial populations.

The scarce environmental exposure data available give some support for the above conclusions:

- reported ambient air levels are low ($\leq 8 \ \mu g/m^3$ for urban areas; $< 200 \ \mu g/m^3$ for heavily industrialized areas);
- phenol has been detected in rain water;
- reported surface water levels are low ($\leq 24 \ \mu g/litre$);
- levels in ground water have only been found at highly contaminated sites.

10.2.2 Toxicity

Based on reported and estimated bioconcentration factors for aquatic organism, phenol is not expected to bioaccumulate significantly. The data base on aquatic toxicity is considered adequate for evaluation. Phenol is toxic to aquatic organisms: the lowest EC_{50} for water organisms is estimated to be 3.1 mg/litre (48-h LC_{50} for *Ceriodaphnia dubia*). The lowest chronic NOEC is estimated to be 0.2 $\mu g/litre$ (8-day LC_1 for *Salmo gairdneri*).

Applying the modified US EPA method, an Environmental Concern Level of 0.02 μg/litre can be derived for water. In general, fish are the most sensitive species and the sensitivities of marine and freshwater organisms are similar. Adequate data on plants and terrestrial organisms are not available.

10.2.3 Evaluation

The scarce exposure data available do not allow any firm conclusions with regard to the degree of risk from phenol to either aquatic or terrestrial ecosystems. However, in view of the derived Environmental Concern Level of phenol for aquatic organisms, it is reasonable to assume that these organisms may be at risk in any surface or sea water subject to phenol contamination, in spite of the rapid degradation of this compound.

11. FURTHER RESEARCH

There is a need for the following items:

a) further investigation of the *in vivo* genotoxicity of phenol;

b) more animal toxicology studies, including 90-day oral and inhalation studies, carcinogenicity bio-assays by the inhalation route, and neurotoxicity and multigeneration reproductive toxicity studies (including evaluation in offspring);

c) assessment of environmental and occupational exposures and evaluation of health effects in occupational populations;

d) further evaluation of the dose-duration-effect relationships, reversibility/persistence and health significance of the reported phenol-induced inhibition of the pupillary response to light;

e) further data on the toxicity of phenol for plants and terrestrial organisms.

12. PREVIOUS EVALUATIONS BY INTERNATIONAL BODIES

The carcinogenic risk of phenol was evaluated in 1989 by the International Agency for Research on Cancer (IARC, 1989). The summary evaluation from the IARC Monograph is reproduced here.

"Exposures

Phenol is a basic feedstock for the production of phenolic resins, bisphenol A, caprolactam, chlorophenols and several alkylphenols and xylenols. Phenol is also used in disinfectants and antiseptics. Occupational exposure to phenol has been reported during its production and use, as well as in the use of phenolic resins in the wood products industry. It has also been detected in automotive exhaust and tobacco smoke.

Experimental carcinogenicity data

Phenol was tested for carcinogenicity by oral administration in drinking-water in one strain of mice and one strain of rats. No treatment-related increase in the incidence of tumours was observed in mice or in female rats. In male rats, an increase in the incidence of leukaemia was observed at the lower dose, but not at the higher dose. Phenol was tested extensively in the two-stage mouse skin model and showed promoting activity.

Human carcinogenicity data

In one case-control study of workers in various wood industries, an increased risk was seen for tumours of the mouth and respiratory tract in association with exposure to phenol; however, the number of cases was small and confounding exposures were inadequately controlled.

Other relevant data

In humans, phenol poisoning can occur after skin absorption, inhalation of vapours or ingestion. Acute local effects are severe tissue irritation and necrosis. At high doses, the most prominent systemic effect is central nervous system depression. Phenol causes irritation, dermatitis, central nervous system effects and liver and kidney toxicity in experimental animals.

Phenol induced micronuclei in female mice and sister chromatid exchange in cultured human cells. It did not inhibit intercellular communication in cultured animal cells. It induced mutation but not DNA damage in cultured animal cells. It did not induce recessive lethal mutation in Drosophila. It had a weak effect in inducing mitotic segregation in *Aspergillus nidulans*. Phenol did not induce mutation in bacteria.

Evaluation

There is inadequate evidence for the carcinogenicity of phenol in humans.

There is inadequate evidence for the carcinogenicity of phenol in experimental animals.

Overall evaluation

Phenol is not classifiable as to its carcinogenicity to humans (Group 3)".

REFERENCES

Abou-el-Makarem MM, Millburn P, Smith RL, & Williams RT (1967) Biliary excretion of foreign compounds. Biochem J, **105**: 1269-1274.

Abraham AJ (1972) A case of carbolic acid gangrene of the thumb. Br J Plast Surg, **25**: 282-284.

ACGIH (1991) Threshold limit values for chemical substances and physical agents and biological exposure indices for 1991-1992. Cincinnati, Ohio, American Conference of Environmental Industrial Hygienists.

Adachi A, Asaka Y, Ozasa M, Sawai N, & Kobayashi T (1987) [Formation of cyanide ion by the reaction of phenol with nitrous acid in waste water.] Eisei Kagaku, **33**(6): 445-448 (in Japanese, with English abstract).

Aelion CM, Swindoll CM, & Pfaender FK (1987) Adaptation to and biodegradation of xenobiotic compounds by microbial communities from a pristine aquifer. Appl Environ Microbiol, **53**(9): 2212-2217.

Agrawal HP (1987) Evaluation of the toxicity of phenol and sodium pentachlorophenate to the snail *Indoplanorbis exustus* (Deshayes). J Anim Morphol Physiol, **34**(1-2): 107-112.

Alabaster JS (1969) Survival of fish in 164 herbicides, insecticides, fungicides, wetting agents and miscellaneous substances. Int Pest Control, **March/April**: 29-35.

Alekseyev VA & Antipin BN (1976) Toxicological characteristics and symptoms of acute phenol poisoning in some freshwater crustaceans and molluscs. Hydrobiol J, **12**(2): 27-33.

Alexander DG & Clarke RMcV (1978) The selection and limitations of phenol as a reference toxicant to detect differences in sensitivity among groups of rainbow trout (*Salmo gairdneri*). Water Res, **12**: 1085-1090.

Al-Sabti K (1985) Frequency of chromosomal aberrations in the rainbow trout, *Salmo gairdneri*, exposed to 5 pollutants. J Fish Biol, **26**(1): 13-20.

American Public Health Association, American Waterworks Association, Water Pollution Control Federation (1985) Standard methods for the examination of water and wastewater - Phenol, 16th ed. Washington, DC, American Public Health Association, pp 556-570.

Amlathe S, Upadhyay S, & Gupta VK (1987) Spectrophotometric determination of trace amounts of phenol in waste water and biological fluids. Analyst, **112**: 1463-1465.

Amoore JE & Hautala E (1983) Odor as an aid to chemical safety; odor thresholds compared with threshold limit values and volatilities for 214 industrial chemicals in air and water dilution. J Appl Toxicol, **3**: 272-290.

Andersen W (1869) Fatal misadventure with carbolic acid. Lancet, **1**: 179.

Anniko M, Hellstrom S, Schmidt SH, & Spandow O (1988) Toxic effects on inner ear of noxious agents passing through the round window membrane. Acta Otolaryngol (Stockholm), **457**(Suppl): 49-56.

Aquino MD, Korol S, Santini P, & Moretton J (1988) Biodegradation of phenolic compounds: I. Improved degradation of phenol and benzoate by indigenous strains of Acinobacter and Pseudomonas. Rev Latinoam Microbiol, 30(3): 283-288.

Aranyi C, O'Shea WJ, Graham JA, & Miller FJ (1986) The effects of inhalation of organic chemical air contaminants on murine lung host defences. Fumdam Appl Toxicol, 6(4): 713-720.

Arvin E, Jensen BK, & Gundersen TA (1991) Biodegradation kinetics of phenol in an aerobic biofilm at low concentrations. Water Sci Technol, 23: 1375-1384.

Atkinson GL & Shupak RC (1989) Acute bronchospasm complicating intercostal nerve block. Anesth Analg, 68: 400-401.

Atkinson R, Darnall KR, Lloyd AC, Winer AM, & Pitts JN Jr (1979) Kinetics and mechanisms of the reactions of the hydroxyl radical with organic compounds int he gas phase. Adv in Photochem, 11: 375.

ATSDR (1989) Toxicological profile for phenol. Atlanta, Georgia, Agency for Toxic Substances and Disease Registry, 111 pp (ATSDR/TP-89/20).

Babich H & Stotzky G (1985) A microbial assay for determining the influence of physicochemical environmental factors on the toxicity of organics: phenol. Arch Environ Contam Toxicol, 14(4): 409-416.

Baird RB, Kuo CL, Shapiro JS, & Yanko WA (1974) The fate of phenolics in waste water - determination by direct injection GLC and Warburg respirometry. Arch Environ Contam Toxicol, 2: 165-169.

Bak F & Widdell F (1986) Anaerobic degradation of phenol and phenol derivatives by *Desulfobacterium phenolicum*, new species. Arch Microbiol, 146(2): 177-180.

Baker MD & Mayfield CI (1980) Microbial and nonbiological decomposition of chlorophenols and phenol in soil. Water Air Soil Pollut, 13: 411-424.

Baker EL, Bertozzi PE, Field PH, Basteyns BJ, & Skinner HG (1978) Phenol poisoning due to contaminated drinking water. Arch Environ Health, 33: 89-94.

Barale R, Marrazzini A, Betti C, Vangelisti F, Loprieno N, & Barrai J (1990) Genotoxicity of two metabolites of benzene: phenol and hydroquinone show strong synergistic effects *in vivo*. Mutat Res, 244: 15-20.

Baranowska-Dutkiewicz B (1981) Skin absorption of phenol from aqueous solutions in men. Int Arch Occup Environ Health, 49: 99-104.

Behl CR, Linn EE, Flynn GL, Pierson CL, Higuchi WI, & Ho NFH (1983a) Permeation of skin and eschar by antiseptics: I. Baseline studies with phenol. J Pharm Sci, 72(4): 391-397.

Behl CR, Linn EE, Flynn GL, Ho NFH, Higuchi WI, & Pierson CL (1983b) Permeation of skin and eschar by antiseptics: II. Influence of controlled burns on the permeation of phenol. J Pharm Sci, 72(4): 397-400.

Bennett IL, James DF, & Golden A (1950) Severe acidosis due to phenol poisoning. Report of two cases. Ann Intern Med, **32**: 324-327.

Benzon HT (1979) Convulsions secondary to intravascular phenol: a hazard of celiac plexus block. Anesth Analg, **58**(2): 150-151.

Beyer J & Frank G (1985) Hydroxylation and conjugation of phenol by the frog, *Rana temporaria*. Xenobiotica, **15**(4): 277-280.

Bergshoeff G (1960) [Determination of phenols in air and water. Part III-A. Instructions for the method using p-nitroaniline and for the method using ultraviolet spectrophotometry.] Delft, The Netherlands, Dutch Organization for Applied Scientific Research (TNO-IMG) (Report No. F765) (in Dutch).

Bernard DP & Salt DB (1982) Carcinogen specificity of gamma-glutamyl transpeptidase-rich foci chemically induced in hamster buccal pouch epithelium. 1st Int. Conf., Boston, Mass., 173-182.

Birge WJ, Black JA, Hudson JE, & Bruser DM ((1979) Embryo larval toxicity test with organic compounds - Aquatic Toxicology. In: Proceedings of the Second Annual Symposium on Aquatic Toxicology, 31 October-1 November 1977. pp 131-147.

Birge WJ, Black JA, & Kuehne RA (1980) Effects of organic compounds on amphibial reproduction. Lexington, Kentucky Water Resources Research Institute, 50 pp.

Blaber LC & Gallagher JP (1971) The facilitory effects of catechol and phenol at the neuromuscular junction of the cat. Neuropharmacology, **10**: 153-159.

Black JA, Birge WJ, Westerman AC, & Francis PC (1983) Comparative aquatic toxicology of aromatic hydrocarbons. Fundam Appl Toxicol, **3**(5): 353-358.

Bohrman JS, Burg JR, Elmore E, Gulati DK, Barfknecht TR, Niemeier RW, Dames BL, Toraason M, & Langenbach R (1988) Interlaboratory studies with the Chinese hamster V79 cell metabolic cooperation assay to detect tumor-promoting agents. Environ Mol Mutagen, **12**(1): 33-51.

Boiano JM (1985) Health hazard evaluation report, University of Pennsylvania, Medical Education Building, Philadelphia, Pennsylvania. Cincinnati, OH, National Institute for Occupational Safety and Health (HETA 84-098-1497; NTIS/PB85-220861).

Bolcsak LE & Nerland DE (1983) Inhibition of erythropoiesis by benzene and benzene metabolites. Toxicol Appl Pharmacol, **69**(3): 363-368.

Bollig FJ & Decker KH (1980) [Phenol resins.] Kunststoffe, **70**: 672 (in German).

Bolt HM (1977) Structural modifications in contraceptive steroids altering their metabolism and toxicity. Arch Toxicol, **39**: 13-19.

Borovik EB & Dmitriev MT (1981) [Principles of air interchange as a basis for the hygienic standardization of the air inside hospitals.] Gig i Sanit, **2**: 22-25 (in Russian).

Boutwell RK & Bosch DK (1959) The tumour promoting action of phenol and related compounds for mouse skin. Cancer Res, **19**: 413-424.

Boutwell RK, Rusch HP, & Bosch DK (1955) The action of phenol and related compounds in tumour formation. Proc Am Assoc Cancer Res, 2: 6-7.

Boutwell RK, Rusch HP, & Booth B (1956) Tumour production by phenol and related compounds. Proc Am Assoc Cancer Res, 2: 96.

Boyd SA (1982) Adsorption of substituted phenols by soil. Soil Sci, 134(5): 337-343.

Bradbury SP, Henry TR, Niemi GJ, Carlson RW, & Snarski VM (1989) Use of respiratory-cardiovascular responses of rainbow trout (*Salmo gairdneri*) in identifying acute toxicity syndromes in fish: Part 3. Polar narcotics. Environ Toxicol Chem, 9(3): 247-262.

Bratzler LJ, Spooner ME, Weatherspoon JB, & Maxey JA (1969) Smoke flavour as related to phenol, carbonyl and acid content of Bologna. J Food Sci, 34: 146-148.

Bray HG, Humphris BG, Thorpe WV, White K, & Wood PB (1952a) Kinetic studies of the metabolism of foreign organic compounds. 3. The conjugation of phenol with glucuronic acid. Biochem J, 52: 416-419.

Bray HG, Humphris BG, Thorpe WV, White K, & Wood PB (1952b) Kinetic studies of the metabolism of foreign organic compounds. 4. The conjugation of phenol with sulfuric acid. Biochem J, 52: 419-423.

Bray HG, Thorpe WV, & White K (1952c) Kinetic studies of the metabolism of foreign organic compounds. 5. A mathematical model expressing the metabolic fate of phenols, benzoic acids and their precursors. Biochem J, 52: 423-430.

Bringmann G & Kuhn R (1959) [Studies on water toxicology using protozoa as test organisms.] Gesundheits-Ing, 80: 239-242 (in German).

Bringmann G & Kuhn R (1977) [Threshold values for the harmful effect of water pollutants on bacteria (*Ps. putida*) and green algae (*Sc. quadricauda*) in the cell reproduction inhibition test.] Z Wasser Abwasser Forsch, 80: 87-98 (in German).

Bringmann G & Kuhn R (1978) [Threshold values for the harmful effect of water pollutants on blue algae (*Microcystis aeruginosa*) and green algae (*Scenedesmus quadricauda*) in the cell reproduction inhibition test.] Vom Wasser, 80: 45-60 (in German).

Bringmann G & Kuhn R (1980) Comparison of the toxicity thresholds of water pollutants to bacteria, algae, and protozoa in the cell multiplication inhibition test. Water Res, 14(3): 231-241.

Bringmann G, Kuhn R, & Winter A (1980) [Determination of the harmful biological action of water endangering substances on protozoa. III. Saprozic flagellates.] Z Wasser Abwasser Forsch, 13(5): 170-173 (in German).

Brodzinsky R & Singh HB (1982) Volatile organic chemicals in the atmosphere: an assessment of available data. Research Triangle Park, North Carolina, US Environmental Protection Agency, Environmental Sciences Research Laboratory (EPA/600/3-83/027a; NTIS PB83-195503).

Brown VM, Jordan DHM, & Tiller BA (1967) The effects of temperature on the acute toxicity of phenol to rainbow trout in hard water. Water Res, 1: 587-594.

Brown VKH, Box VL, & Simpson BJ (1975) Decontamination procedures for skin exposed to phenolic substances. Arch Environ Health, **30**: 1-6.

Bruce RM, Santodonato J, & Neal MW (1987) Summary review of the health effects associated with phenol. Toxicol Ind Health, **3**(4): 535-568.

Budavari S, O'Neil MJ, Smith A, & Heckelman PE ed. (1989) The Merck index. Rahway, New Jersey, Merck & Co., Inc., p 1150.

Buikema AL, McGinness MJ, & Cairns J (1979) Phenolics in aquatic ecosystems. A selected review of recent literature. Mar Environ Res, **2**(2): 87-181.

Bulsiewicz H (1977) The influence of phenol on chromosomes of mice in the process of spermatogenesis. Folia Morphol (Warsaw), **36**: 13-22.

Cairns J, Messenger DJ, & Calhoun WF (1976) Invertebrate response to thermal shock following exposure to acutely sub-lethal concentrations of chemicals. Arch Hydrobiol, **77**(2): 164-175.

Cairns J, Buikema AL Jr, Heath AG, & Parker BC (1978) Effects of temperature on aquatic organism sensitivity to selected chemicals. Blackburg, Virginia, Virginia Water Resources Research Center (Bulletin 106).

Call DJ, Brooke LT, & Lu PY (1980) Uptake, elimination, and metabolism of three phenols by Fathead minnow. Arch Environ Contam Toxicol, **9**(6): 699-714.

Campbell NRC, Van Loon JA, & Weinshilboum RM (1987) Human liver phenol sulfotransferase: assay conditions, biochemical properties and partial purification of isozymes of the thermostable form. Biochem Pharmacol, **36**(9): 1435-1446.

Capel ID, French MR, Millburn P, Smith RL, & Williams RT (1972a) Species variations in phenol metabolism. Biochem J, **127**: 25-26.

Capel ID, French MR, Millburn P, Smith RL, & Williams RT (1972b) The fate of ^{14}C phenol in various species. Xenobiotica, **2**: 25-34.

Capel ID, Millburn P, & Williams RT (1974) Monophenyl phosphate, a new conjugate of phenol in the cat. Biochem Soc Trans, **2**: 305-306.

Cassidy MK & Houston JB (1980) *In vivo* assessment of extrahepatic conjugative metabolism in first pass effects using the model compound phenol. J Pharm Pharmacol, **32**(5): 57-59.

Cassidy MK & Houston JB (1984) *In vivo* capacity of hepatic and extrahepatic enzymes to conjugate phenol. Drug Metab Dispos, **12**: 619-624.

Chagnon N & Hlohowskyj I (1989) Effects of phenol exposure on the thermal tolerance ability of the central stoneroller minnow. Bull Environ Contam Toxicol, **42**(4): 614-619.

Chemfacts (1978-1981) Chemical Data Sources - Chemfacts. Publications for the Netherlands, Belgium, West-Germany, France, Great Britain, Spain, Italy, Portugal and Scandinavia.

Chen TH, Kavanagh TJ, Chang CC, & Trosko JE (1984) Inhibition of metabolic cooperation in Chinese hamster V79 cells by various organic solvents and simple compounds. Cell Biol Toxicol, 1(1): 155-171.

Chesney RH, Sollitti P, & Rubin HE (1985) Incorporation of phenol carbon at trace concentrations by phenol-mineralizing microorganisms in fresh water. Appl Environ Microbiol, 49(1): 15-18.

CID-TNO (1984) Databases ECDIN and PROMT.

Ciranni R, Barale R, Ghelardini G, & Loprieno N (1988a) Benzene and the genotoxicity of its metabolites. I. Transplacental activity in mouse fetuses and in their dams. Mutat Res, 208: 61-67.

Ciranni R, Barale R, Ghelardini G, & Loprieno N (1988b) Benzene and the genotoxicity of its metabolites. II. The effect of the route of administration on the micronuclei and bone marrow depression in mouse bone marrow cells. Mutat Res, 209: 23-28.

Coan ML, Baggs RB, & Bosmann HB (1982) Demonstration of direct toxicity of phenol on kidney. Chem Pathol Pharmacol Res Commun, 36(2): 229-239.

Colvin RJ & Rozich AR (1986) Phenol growth kinetics of heterogenous populations in a two-stage continuous culture system. Water Pollut Control Fed, 58(4): 326-332.

Conning DM & Hayes MJ (1970) The dermal toxicity of phenol, and investigation of the most effective first-aid measures. Br J Ind Med, 27: 155-159.

Cooper IS, Hirose T, Matsuoka S, Roth D, Waltz JM, & Ericsson AD (1965) Specific neurotic perfusion: a new approach to selected cases of pain and spasticity. Neurology, 15: 985.

Cotruvo JA, Simmon VF, & Sponggord RJ (1977) Investigation of mutagenic effects of products of ozonation reactions in water. Ann NY Acad Sci, 298: 124-140.

Crebelli R, Conti G, & Carere A (1987) On the mechanism of mitotic segregation induction in *Aspergillus nidulans* by benzene hydroxy metabolites. Mutagenesis, 2(3): 235-238.

Cronin TD & Brauer RO (1949) Death due to phenol contained in Foille. J Am Med Assoc, 139: 777-779.

Curtis C, Lima A, Lozano SJ, & Veith GD (1982) Philadelphia, Pennsylvania, American Society for Testing and Materials, p 170 (ASTM Special Technical Publication No. 766).

Dalela RC, Rani S, & Verna SR (1980) Influence of pH on the toxicity of phenol and its 2 derivatives pentachlorophenol and dinitrophenol to some fresh water teleosts. Acta Hydrochim Hydrobiol, 8(6): 623-628.

Dalin NM & Kristofferson R (1974) Physiological effects of a sublethal concentration of inhaled phenol on the rat. Ann Zool Fenn, 11: 193-199.

Dangé AD (1986) Metabolic effects of naphthalene, toluene or phenol intoxication in the cichlid fish tilapia, *Oreochromis mossambicus*: changes in aminotransferase activities. Environ Pollut, A42: 311-323.

Dauble DD, Carlile DW, & Hanf RW Jr (1986) Bioaccumulation of fossil fuel components during single-compound and complex-mixture exposures of *Daphnia magna*. Bull Environ Contam Toxicol, **37**: 125-132.

Daum F, Cohen MJ, & McNamara H (1976) Experimental toxicological studies on a phenol detergent associated with neonatal hyperbilirubinemia. J Pediatr, **89**: 853-854.

Davis HC & Hidu H (1969) Effects of pesticides on embryonic development of clams and oysters and on survival and growth of the larvae. US Dept. Inter., Fish Wildl. Bull., **67**: 393.

De Ceaurriz JC, Micillino JC, Bonnet P, & Guenier JP (1981) Sensory irritation caused by various industrial airborne chemicals. Toxicol Lett, **9**: 137-143.

De Graeve GM, Geiger WL, Meyer JS, & Bergman HL (1980) Acute and embryo-larval toxicity of phenolic compounds to aquatic biota. Arch Environ Contam Toxicol, **9**: 557-568.

De Groote V & Lambotte C (1960) A case of fatal phenol poisoning in a newborn infant. Ann Méd Lég, **40**: 288-290.

Deichmann WB (1944) Phenol studies. V. The distribution, detoxification, and excretion of phenol in the mammalian body. Arch Biochem, **2**: 345-355.

Deichmann WB (1949) Local and systemic effects following skin contact with phenol: a review of the literature. J Ind Hyg Toxicol, **31**: 146-154.

Deichmann WB & Keplinger ML (1963) Phenols and phenolic compounds. In: Patty FA ed. Industrial hygiene and toxicology. New York, Interscience Publishers, pp 2567-2627.

Deichmann WB & Oesper P (1940) Ingestion of phenol - effects of the albino rat. Ind Med, **9**: 296.

Deichmann WB & Schafer LJ (1942) Phenols studies - I: Review of the literature. II: Quantitative spectrophotometric estimation of free and conjugated phenol in tissues and fluids. III: Phenol content of normal human tissues and fluids. Am J Clin Pathol, **12**: 129-143.

Deichmann WB & Witherup S (1944) Phenol studies - VI: The acute and comparative toxicity of phenol and o-, m-, and p-cresols for experimental animals. J Pharmacol Exp Ther, **80**: 233-240.

Deichmann WB, Kitzmiller KV, & Witherup S (1944) Phenols studies - VII: Chronic phenol poisoning, with special reference to the effects upon experimental animals of the inhalation of phenol vapour. Am J Clin Pathol, **14**: 273-277.

Deichmann WB, Miller T, & Roberts JB (1950) Local and systemic effects following application of dilute solutions of phenol in water and in camphor-liquid petrolatum on the skin of animals. Arch Ind Hyg Occup Med, **2**: 454-461.

Deichmann WB, Witherup S, & Dierker M (1952) Phenol studies - XII: The percutaneous and alimentary absorption of phenol by rabbits with recommendations for the removal of phenol from the alimentary tract or skin of persons suffering exposure. J Pharmacol Exp Ther, **105**: 265-272.

Delfino JJ & Dube DJ (1976) Persistent contamination of ground water by phenol. J Environ Sci Health, pp 345-355.

DeMeio RH & Arnolt RI (1944) Phenol conjugation - II: The conjugation by rat and cat tissues *in vitro*. J Biol Chem, **156**: 577-583.

Dement J, Wallingford K, & Zumwalde R (1973) Industrial hygiene survey of Owens-Corning Fiberglas, Kansas, City, Kansas. Cincinnati, Ohio, National Institute for Occupational Safety and Health (Report No. IW 35.16).

Demerec M, Bertani G, & Flint J (1951) A survey of chemicals for mutagenic action on *E. coli*. Am Nat, **85**: 119-136.

Den Boeft J, Kruiswijk FJ, & Schulting FL (1984) [Air pollution by combustion of solid fuels.] The Hague, Ministry of Housing, Physical Planning and Environment (Publication Lucht No. 37) (in Dutch).

De Neer JW, Seinen W, & Hermens JLM (1988) Growth of *Daphnia magna* exposed to mixtures of chemicals with diverse modes of action. Ecotoxicol Environ Saf, **15**(1): 72-77.

Dimitriev MT & Mischchikhin VA (1983) [Gas chromatographic determination of phenol in air.] Gig i Sanit, **5**: 42-44 (in Russian).

Dive O & Leclerc H (1977) Utilisation du protozoaire cilie colpidum campylum pour la meure de la toxicité et de l'accumulation des micropollutants: analyse critique et applications. Environ Pollut, **14**: 169-186.

Doan HMcK, Keith L, & Shennan AT (1979) Phenol and neonatal jaundice. Pediatrics, **64**(3): 324-325.

Dobbins DC, Thornton-Manning JR, Jones DD, & Federle TW (1987) Mineralization potential for phenol in subsurface soils. J Environ Qual, **16**(1): 54-58.

Docter HJ & Zielhuis RL (1967) Phenol excretion as a measure of benzene exposure. Ann Occup Hyg, **10**: 317-326.

Dosemeci M, Blair A, Stewart, PA, Chandeler J, & Trush MA (1991) Mortality among industrial workers exposed to phenol. Epidemiology, **2**, 188-193.

Dow Chemical Company (1976) References and literature review pertaining to toxicological properties of phenol. Dow Chemical Company, Toxicological Research Laboratory (unpublished manuscript).

Dowden BF & Benett HJ (1965) Toxicity of selected chemicals to certain animals. J Water Pollut Control Fed, **37**: 1308-1316.

Dreibelbis WG & Hawthorne AR (1985) Worker exposure potential at coal conversion facilities. Oak Ridge, Tennessee, Oak Ridge National Laboratory (NTIS/DE85006822).

Dreisbach RH (1983) Handbook of poisoning: Prevention, diagnosis & treatment, 11th ed. Los Altos, California, Lange Medical Publications, pp 401-405.

Dumpert K (1987) Embryotoxic effects of environmental chemicals: tests with the South African clawed toad (*Xenopus laevis*). Ecotoxicol Environ Saf, **13**(3): 324-338.

Dutka BJ & Kwann KK (1981) Comparison of three microbial toxicity screening tests with the Microtox test. Bull Environ Contam Toxicol, **27**: 753-757.

Dutka BJ, Nyholm N, & Peterson J (1983) Comparison of several microbiological toxicity screening tests. Water Res, **17**(10): 1363-1368.

Duverneuil G & Ravier E (1962) Toxicité suraiguë du phénol par voie transcutanée. Arch Mal Prof, **23**: 830-833.

Eastmond DA, Smith MT, Ruzo LG, & Ross D (1986) Metabolic activation of phenol by human myeloperoxidase and horseradish peroxidase. Mol Pharmacol, **30**(6): 674-679.

Eastmond DA, French RG, Ross D, & Smith MT (1987a) Metabolic activation of 1-naphthol and phenol by a simple superoxide-generating system and human leukocytes. Chem-Biol Interact, **63**(1): 47-62.

Eastmond DA, Smith MT, & Irons RD (1987b) An interaction of benzene metabolites reproduces the myelotoxicity observed with benzene exposure. Toxicol Appl Pharmacol, **91**: 85-95.

Edwards VT, Jones BC, & Hudson DH (1986) A comparison of the metabolic fate of phenol, phenyl glucoside and phenyl 6-o-malonyl-glucoside in the rat. Xenobiotica, **16**: 801-807.

Ehrlich GG, Goelitz DF, & Godsy EM (1982) Degradation of phenolic contaminants in ground water by anaerobic bacteria: St. Louis Park, MN. Ground Water, **20**: 703-710.

EIFAC (European Inland Fisheries Advisory Commission) (1972) Water quality criteria for European freshwater fish. Report on monohydric phenols and inland fisheries. Rome, Food and Agriculture Organization of the United Nations (Technical Paper No. 15).

Epler JL, Rao TK, & Guerin MR (1979) Evaluation of feasibility of mutagenic listing of shale oil products and effluents. Environ Health Perspect, **30**: 179-184.

Erben R (1982) The effect of phenol and its accumulation on snails *Physa fontinalis* and *Radix peregra mull* (Pulmonata). Period Biol, **84**(4): 403-410.

Erben R, Munjko J, & Jovanovic V (1983) Influence of phenol accumulation on the species *Polycelis felina daly* (Turbellaria). Biologica (Bratislava), **38**: 283-288.

Erexson GL, Wilmer JL, & Kligerman AD (1985) Sister chromatid exchange induction in human lymphocytes exposed to benzene and its metabolites *in vitro*. Cancer Res, **45**(6): 2471-2477.

Ernst MR, Klesmer R, Huebner RA, & Martin JE (1961) Susceptibility of cats to phenol. J Am Vet Med Assoc, **138**: 197-199.

Ersek RA (1991) Comparative study of dermabrasion, phenol peel, and acetic acid peel. Anesth Plast Surg, **15**: 241-243.

Evans SJ (1952) Acute phenol poisoning. Br J Ind Med, **9**: 227-229.

Farquharson ME, Gage JC, & Northover J (1958) The biological action of chlorophenols. Br J Pharmacol, **13**: 20-24.

Federle TW (1988) Mineralization of monosubstituted aromatic compounds in unsaturated and saturated subsurface soils. Can J Microbiol, 34(9): 1037-1042.

Fedorak PM & Hrudey SE (1986) Nutrient requirements for the methanogenic degradation of phenol and p-cresol in anaerobic draw and feed cultures. Water Res, 20(7): 929-934.

Fedorak PM, Roberts DJ, & Hrudey SE (1986) The effects of cyanide on the methanogenic degradation of phenolic compounds. Water Res, 20(10): 1315-1320.

Feldman RJ & Maibach HI (1970) Absorption of some organic compounds through the skin in man. J Invest Dermatol, 54: 399-404.

Fielding M, Gibson TM, James HA, et al (1981) Organic micropollutants in drinking water. Medmenham, UK, Water Research Centre (Report No. TR-159).

Fishbeck WA, Langner RR, & Kociba RJ (1975) Elevated urinary phenol levels not related to benzene exposure. Am Ind Hyg Assoc J, 36(11): 820-824.

Flerov BA (1979) Physiological mechanisms of the action of toxic substances and adaptation of aquatic animals to them. Hydrobiol J, 13(4): 70-74.

Flickinger CW (1976) The benzenediols: catechol, resorcinol and hydroquinone - a review of the industrial toxicology and current industrial exposure limits. Am Ind Hyg Assoc J, 37: 596-606.

Flynn GL & Yalkowsky SH (1972) Correlation and prediction of mass transport across membranes. I. Influence of alkyl chain length on flux-determining properties of barrier and diffusant. J Pharm Sci, 61: 838-852.

Fogels A & Sprague JB (1977) Comparative short-term tolerance of zebrafish, flagfish, and rainbow trout to five poisons including potential reference toxicants. Water Res, 11(9): 811-817.

Forrest F & Ramage DTO (1987) Cardiac dysrhythmia after subtrigonal phenol. Anaesthesia, 42(7): 777-778.

Foxall PJD, Bending MR, Gartland KPR, & Nicholson JK (1991) Acute renal failure following accidental cutaneous absorption of phenol. In: Bach PH ed. Nephrotoxicity: Mechanisms, early diagnosis and therapeutic management (Fourth International Symposium on Nephrotoxicity, Guildford, UK, 1989). New York, Basel, Marcel Dekker, Inc., pp 55-59.

Franz TJ (1975) Percutaneous absorption. On the relevance of *in vitro* data. J Invest Dermatol, 64: 190-195.

Freitag D, Ballhorn L, Geyer H, & Korte F (1985) Environmental hazard profile of organic chemicals. An experimental method for the assessment of the behaviour of organic chemicals in the ecosphere by means of simple laboratory tests with ^{14}C-labelled chemicals. Chemosphere, 14(10): 1589-1616.

French MR, Bababunmi EA, Golding RR, Bassir O, Caldwell J, Smith RL, & Williams RT (1974) The conjugation of phenol, benzoic acid, 1-naphthylacetic acid and sulphadimethoxine in the lion, civet and genet. FEBS Lett, 46(1): 134-137.

Fujii M (1978) Environmental monitoring of chemicals - Environmental survey report of FY 1977. Tokyo, Environment Agency Japan, Department of Environmental Health.

Gad-El Karim MM, Ramanujam VMS, Ahmed AE, & Legator MS (1965) Benzene mycloclastogenicity: A function of its metabolism. Am J Ind Med, 7: 475-484.

Gad-El Karim MM, Ramanujam VMS, & Legator MS (1966) Correlation between the induction of micronuclei in bone marrow by benzene exposure and the excretion of metabolites in urine of CD-1 mice. Toxicol Appl Pharmacol, **85**: 464-477.

Garberg P, Akerblom EL, & Bolcsfoldi G (1988) Evaluation of a genotoxicity test measuring DNA-strand breaks in mouse lymphoma cells by alkaline unwinding and hydroxyapatite elution. Mutat Res, 203: 155-176.

Garton GA & Williams RT (1949) Studies in detoxification. 26. The fate of phenol, phenyl sulfuric acids and phenylglucuronide in the rabbit in relation to the metabolism of benzene. Biochem J, **45**: 158-163.

Gbodi TA & Oehme FW (1978) The fate of phenol, o-phenylphenol and diisophenol in rats. Toxicol Appl Pharmacol, **45**: 219-362.

Gersich FM, Blanchard FA, Applegath SL, & Park CN (1986) The precision of daphnid (*Daphnia magna* Straus, 1820) static acute toxicity tests. Arch Environ Contam Toxicol, **15**: 741-749.

Gibb C, Glover V, & Sandler M (1987) *In vitro* inhibition of phenol sulphotransferase by food and drink constituents. Biochem Pharmacol, **36**(14): 2325-2330.

Gibson II (1987) Phenol block in the treatment of spasticity. Gerontology, **33**(5): 327-330.

Gilbert P, Rondelet J, Poncelet F, & Mercier M (1980) Mutagenicity of p-nitrosophenol. Food Cosmet Toxicol, **18**: 523-525.

Gilli G, Palin L, & Scursatore E (1980) [Exposition to phenol.] Arch Sci Med, **137**: 41-44 (in Italian).

Gluth G & Hanke W (1983) The effect of temperature on physiological changes in Carp *Cyprinus carpio* L. induced by phenol. Ecotoxicol Environ Saf, 7(4): 373-389.

Gocke E, King MT, Eckhardt K, & Wild D (1981) Mutagenicity of cosmetics, ingredients licensed by the European Communities. Mutat Res, **90**: 91-109.

Görge G, Beyer J, & Urich K (1987) Excretion and metabolism of phenol, 4-nitrophenol and 2-methylphenol by the frogs *Rana temporaria* and *Xenopus laevis*. Xenobiotica, **17**(11): 1293-1298.

Green DWJ, Williams KA, & Pascoe D (1985) Studies on the acute toxicity of pollutants to freshwater macroinvertebrates: 2. Phenol. Arch Hydrobiol, **103**(1): 75-82.

Green DWJ, Williams KA, Hughes DRL, Shaik GAR, & Pascoe D (1988) Toxicity of phenol to *Asellus aquaticus* (L.): effects of temperature and episodic exposure. Water Res, **22**(2): 225-232.

Greenlee WF, Gross EA, & Irons RW (1981) Relationship between benzene toxicity and the disposition of ^{14}C-labelled benzene metabolites in the rat. Chem-Biol Interact, **33**: 285-299.

Groenen PJ (1978) [Components of tobacco smoke. Nature and quantity; potential influence on health.] Zeist, The Netherlands, CIVO-TNO Institute (Report No. R/5787) (in Dutch).

Gspan P, Jersic A, & Cadez E (1984) [Relationship between phenol concentration in urine and work environment.] Staub-Reinhalt Luft, **44**: 314-316 (in German).

Gunter BJ (1987) Health hazard evaluation report: CF & I Steel, Pueblo, Colorado. Cincinnati, Ohio, National Institute for Occupational Safety and Health (HETA 86-034-1712; NTIS/PB87-117016).

Gupta S & Dalela RC (1987) Kidney damage in *Notopterus notopterus* (Pallas) following exposure to phenolic compounds. J Environ Biol, **8**(2): 167-172.

Gupta PK & Rao PS (1982) Toxicity of phenol, pentachlorophenol and sodium pentachlorophenolate to a freshwater pulmonate snail *Lymnaea acuminata* (Lamarck). Arch Hydrobiol, **94**: 210-217.

Gupta PK, Mujumdar VS, Rao PS, & Durve VS (1982a) Toxicity of phenol, pentachlorophenol and sodium pentachlorophenolate to a freshwater teleost *Lebistes reticulatus* (Peters). Acta Hydrochim Hydrobiol, **10**(2): 177-181.

Gupta S, Verma Sr, & Saxena PK (1982b) Toxicity of phenolic compounds in relation to the size of a freshwater fish, *Notopterus notopterus* (Pallas). Ecotoxicol Environ Saf, **6**: 433-438.

Gupta S, Dalela RC, & Saxena PK (1983a) Influence of dissolved oxygen levels on acute toxicity of phenolic compounds to a freshwater teleost, *Notopterus notopterus* (Pallas). W.A.s. Pol., **19**(3): 223-228.

Gupta S, Dalela RC, & Saxena PK (1983b) Influence of temperature on the toxicity of phenol and its chloro and nitro-derivates to the fish *Notopterus notopterus* (Pallas). Acta Hydrochim Hydrobiol, **11**(2): 187-192.

Gurujeyalashmi S & Oriel P (1989) Isolation of phenol-degrading Bacillus stearothermophilus and partial characterization of the phenol hydroxylase. Appl Environ Microbiol, **55**(2): 500-502.

Haddad LM, Dimend KA, & Schweistris JE (1979) Phenol poisoning. JACEP, **8**(7): 267-269.

Hadorn E & Niggli N (1946) Mutations in Drosophila after chemical treatment of gonads *in vitro*. Nature (Lond), **157**: 162-163.

Hagemann R, Virelizier H, & Gaudin D (1978) Analyse de composés organiques dans une atmosphère urbaine. Analusis, 6(9): 401-405.

Hardy JT, Dauble DD, & Felice LJ (1985) Aquatic fate of synfuel residuals: bioaccumulation of aniline and phenol by the freshwater phytoplankton *Scenedesmus quadricauda*. Environ Toxicol Chem, **4**(1): 29-35.

Haworth S, Lawlor T, Mortelmans K, Speck W, & Zeiger E (1983) Salmonella mutagenicity test results for 250 chemicals. Environ Mutagen, 1(Suppl): 3-142.

Healy JB Jr & Young LY (1979) Anaerobic biodegradation of eleven aromatic compounds to methane. Appl Environ Microbiol, 28: 84-89.

Heikkilä P, Hämeilä M, Pyy L, & Raunu P (1987) Exposure to creosote in the impregnation and handling of wood. Scand J Work Environ Health, 13: 431-437.

Heller VG & Pursell L (1938) Phenol-contaminated waters and their physiological action. J Pharmacol Exp Ther, 63: 99-107.

Henschler D ed. (1975) [Phenol.] In: [Analytical methods for the testing of working media harmful to health. Volume 1: Air analyses.] Weinheim, Verlag Chemie (in German).

Herbert DWM (1962) The toxicity to rainbow trout of spent still liquors from the distillation of coal. Ann Appl Biol, 50: 755-777.

Hermens J (1984) Joint effects of a mixture of 14 chemicals on mortality and inhibition of reproduction of *Daphnia magna*. (Thesis).

Hickman GT & Novak JT (1989) Relationship between subsurface biodegradation rates and microbial density. Environ Sci Technol, 23(5): 524-532.

Hinkel GK & Kintzel HW (1968) Phenol poisoning of a newborn through skin resorption. Dtsch Gesundh, 23: 2420-2422.

Hoffmann KH & Vogt U (1988) [Degradation of phenol by yeasts in the presence of *n*-hexadecane under growth conditions in a stirred reactor.] Zent.bl Mikrobiol, *143*(2): 87-91 (in German, with English abstract).

Hogg SI, Curtis CG, Upshall DG, & Powell GM (1981) Conjugation of phenol by rat lung. Biochem Pharmacol, 30(12): 1551-1555.

Holcombe GW, Phipps GL, & Fiandt JT (1982) Effects of phenol, 2,4-dimethylphenol, 2,4-dichlorophenol and pentachlorophenol on embryo, larval and early-juvenile fathead minnows (*Pimephales promelas*). Arch Environ Contam Toxicol, 11: 73-78.

Holcombe GW, Phipps GL, Sulaiman AH, & Hoffman AD (1987) Simultaneous multiple species testing: acute toxicity of 13 chemicals to 12 diverse freshwater amphibian, fish, and invertebrate families. Arch Environ Contam Toxicol, *16*(6): 697-710.

Hoshika Y & Muto G (1979) Sensitive gas chromatographic determination of phenols as bromophenols using electron capture detection. J Chromatogr, 179: 105-111.

Hoshika Y & Muto G (1980) Gas-liquid-solid chromatographic determination of phenols in air using Tenax-GC and alkaline precolumns. J Chromatogr, 187: 277-284.

Hoshino M & Akimoto H (1978) Photochemical oxidation of benzene, toluene and ethylbenzene initiated by OH radicals in the gas phase. Bull Chem Soc Jpn, 51: 718.

Houston JB & Cassidy MK (1982) Sites of phenol conjugation in the rat in sulfate metabolism and sulfate conjugation. In: Proceedings of an international workshop held at Noordwijkerhout, The Netherlands, 20-23 September 1981. Taylor & Francis, pp 270-278.

Howard PH (1989) Handbook of environmental fate and exposure data for organic chemicals. Chelsea, Michigan, Lewis Publishers, vol 1, pp 468-476.

Hsieh GC, Sharma RP, Parker RDR, & Coulombe RA Jr (1992) Immunological and neurobiochemical alterations induced by repeated oral exposure of phenol in mice. Eur J Pharmacol - Environ Toxicol Pharmacol Sect, **228**: 107-114.

Hubble BR, Stetter JR, Gebert E, Harkness JBL, & Flotard RD (1981) Experimental measurements from residential wood-burning stoves. Proceedings of the International Conference on Residential Solid Fuels: Environmental Impacts and Solutions, John A. Cooper & Dorothy Malek.

Huq AS, Ho NFH, Husari N, Flynn GL, Jetzer WE, & Condie L (1986) Permeation of water contaminative phenols through hairless mouse skin. Arch Environ Contam Toxicol, **15**(5): 557-566.

Hwang HM, Hodson RE, & Lee RF (1986) Degradation of phenol and chlorophenols by sunlight and microbes in estuarine water. Environ Sci Technol, **20**(10): 1002-1007.

Hwang HM, Hodson RE, & Lewis DL (1989) Assessing interactions of organic compounds during biodegradation of complex waste mixtures by naturally occurring bacterial assemblages. Environ Toxicol Chem, **8**(3): 209-214.

IARC (1989) Phenol. In: Some organic solvents, resin monomers and related compounds, pigments and occupational exposures in paint manufacture and painting. Lyon, International Agency for Research on Cancer, pp 263-287 (IARC Monographs on the Evaluation of Carcinogenic Risks to Humans, Volume 47).

Ikeda M & Ohtsuji H (1969) Hippuric acid, phenol and trichloroacetic acid levels in the urine of Japanese subjects with no known exposure to organic solvents. Br J Ind Med, **26**: 162-164.

Illing HPA & House ESA (1980) Binding of (^{14}C) phenol to rat liver high-speed supernatant. Biochem Soc Trans, **8**(1): 117-118.

Isaacson PJ & Frink GR (1984) Nonreversible sorption of phenolic compounds by sediment fractions: the role of sediment organic matter. Environ Sci Technol, **18**: 43-48.

Itoh M (1982) Sensitisation potency of some phenolic compounds. J Dermatol, **9**(3): 223-283.

Ivett JL, Brown BM, Rodgers C, Anderson BE, Resnick MA, & Zeiger E (1989) Chromosomal aberrations and sister chromatid exchange tests in Chinese hamster ovary cells *in vitro*. IV. Results with 15 chemicals. Environ Mol Mutagen, **14**: 165-187.

Jansson T, Curvall M, Hedin A, & Enzell CR (1986) *In vitro* studies of biological effects of cigarette smoke condensate. II. Induction of sister chromatid exchanges in human lymphocytes by weakly acidic, semivolatile constituents. Mutat Res, **169**: 129-139.

Jarvis SN, Straube RC, Williams ALJ, & Bartlett CLR (1985) Illness associated with contamination of drinking water supplies with phenol. Br Med J, **290**: 1800-1802.

Jelinek R, Peterka M, & Rychter Z (1985) Chick embryotoxicity screening test - 130 substances tested. Indian J Exp Biol, **23**: 588-595.

Jergil B, Schelin C, & Tunek A (1982) Covalent binding of metabolically activated hydrocarbons to specific microsomal proteins. Adv Exp Med Biol, **136A**: 341-348.

Jetzer WE, Huq AS, Ho NFH, Flynn GL, Duraiswamy N, & Condie L (1986) Permeation of mouse skin and silicone rubber membranes by phenols: relationship to *in vitro* partitioning. J Pharm Sci, **75**(11): 1098-1103.

Jetzer WE, Hou SY, Huq AS, Duraiswamy N, Ho NFH, & Flynn GL (1988) Temperature dependency of skin permeation of waterborne organic compounds. Pharm Acta Helv, **63**(7): 197-201.

Jones-Price C, Kimmel CA, Ledoux TA, Reel JR, Fisher PW, Langhoff-Paschke L, & Marr MC (1983a) Final study report - Teratologic evaluation of phenol (CAS No. 108-95-2) in CD rats (NTP Study No. TER-81-104). Springfield, Virginia, National Technical Information Service (NTIS/PB83-247726).

Jones-Price C, Kimmel CA, Ledoux TA, Reel JR, Langhoff-Paschke L, & Marr MC (1983b) Final study report - Teratologic evaluation of phenol (CAS No. 108-95-2) in CD-1 rats (NTP Study No. TER-80-129). Springfield, Virginia, National Technical Information Service (NTIS/ PB85-104461.

Judis J (1982) Binding of selected phenol derivatives to human serum proteins. J Pharm Sci, **71**(110): 1145-1147.

Jünke I & Lüdemann D (1978) [Results of the testing of 200 chemical compounds for acute toxicity for fish by the golden orfe test.] Z Wasser Abwasser Forsch, **11**(5): 161-164 (in German).

Kao J, Bridges JW, & Faulkner JK (1979) Metabolism of ^{14}C phenol by sheep, pig and rat. Xenobiotica, **9**(3): 141-147.

Kasokat T, Nagel R, & Urich K (1987) The metabolism of phenol and substituted phenols in zebra fish. Xenobiotica, **17**(10): 1215-1222.

Katz M ed. (1977) Tentative method of analysis for determination of phenolic compounds in the atmosphere with the 4-amino antipyrine method - Methods of air sampling and analysis, 2nd ed. Washington, DC, American Public Health Association, pp 324-332.

Kauppinen TP, Partanen TJ, Nurminen MM, Nickels JI, Hernberg, SG, Hakulinen TR, Pukkala EI, & Savonen ET (1986) Respiratory cancers and chemical exposures in the wood industry: a nested case-control study. Br J Ind Med, **43**(2): 84-90.

Kavlock RJ (1990) Structure-activity relationships in the developmental toxicity of substituted phenols: *In vivo* effects. Teratology, **41**: 43-59.

Keen R & Baillod CR (1985) Toxicity to Daphnia of the end products of wet oxidation of phenol and substituted phenols. Water Res, **19**(6): 767-772.

Key PB & Scott GI (1986) Lethal and sublethal effects of chlorine, phenol, and chlorine-phenol mixtures on the mud crab *Panopeus herbstii*. Environ Health Perspect, **69**: 307-312.

Khalili AA & Betts HB (1967) Peripheral nerve block with phenol in the management of spasticity. J Am Med Assoc, **200**: 1155.

Kifune I (1979) [Determination of micro amounts of phenol and cresols in the atmosphere by using a filter impregnated with a solvent of alkali and glycerol.] Bunseki Kagaku, **28**(11): 638-642 (in Japanese).

Kirk RE & Othmer DF (1980) Encyclopedia of chemical toxicology, 3rd ed. New York, John Wiley and Sons, vol 17, pp 373-379.

Kligman AM (1966) The identification of contact allergens by human assays. III. The maximization test: A procedure for screening and rating contact sensitizers. J Invest Dermatol, **47**(5): 393-409.

Knapik Z, Hanczyc H, Lubczynska-Kowalska W, Menzel-Lipinska M, Cader J, Paradowski L, & Borowka Z (1980) Use of subclinical forms of the toxic action of phenol. Z Gesamte Hyg Grenzgeb, **26**(8): 585-587.

Knezovich JP, Hirabayashi JM, Bishop DJ, & Harrison FL (1988) The influence of different soil types on the fate of phenol and its biodegradation products. Chemosphere, **17**(11): 2199-2206.

Knoevenagel K & Himmelreich R (1976) Degradation of compounds containing carbon atoms by photooxidation in the presence of water. Arch Environ Contam Toxicol, **4**: 324-333.

Knoll G & Winter J (1987) Anaerobic degradation of phenol in sewage sludge: benzoate formation from phenol and carbon dioxide in the presence of hydrogen. Appl Microbiol Biotechnol, **25**(4): 384-391.

Kobayashi K & Akitake K (1975) Metabolism of chlorophenols in fish. VIV Absorption, and excretion of phenol by goldfish. Bull Jpn Soc Sci Fish, **41**: 1271-1276.

Kobayashi K, Akitake H, & Kimura S (1976) Studies on the metabolism of chlorophenols in fish. VI Turnover of absorbed phenol in goldfish. Bull Jpn Soc Sci Fish, **42**: 45-50.

Kobayashi K, Oshima Y, Hamada S, & Taguchi C (1987) Induction of phenol-sulfate conjugating activity by exposure to phenols and duration of its induced activity in short-necked clam. Bull Jpn Soc Sci Fish, **53**(11): 2073-2076.

Koike N, Haga S, Ubukata N, Sakurai M, Shimizu H, & Sato A (1988) [Mutagenicity of benzene metabolites by fluctuation test.] Jpn. J. Ind. Health **30**(6): 475-480 (in Japanese, with English abstract).

Koop DR, Laethem CL, & Schnier GG (1989) Identification of ethanol-inducible P450 isozyme 3a (P450IIE1) as a benzene and phenol hydroxylase. Toxicol Appl Pharmacol, **98**: 278-288.

Kordylewska A (1980) Developmental disturbances of *Limnaea stagnalis* due to phenol intoxication. Acta Biol (Cracow) Ser Zool, **22**(1): 105-130.

Koster HJ, Halsema J, Scholtens E, Knippers M, & Mulder GJ (1981) Dose-dependent shifts in the sulfation and glucuronidation of phenolic compounds in the rat *in vivo* and in isolated hepatocytes. The role of saturation of phenolsulfotransferase. Biochem Pharmacol, **30**(18): 2569-2575.

Koster HJ (1982) Sulfuration and glucuronidation in the rat *in vivo* and *in vitro*. The balance between two enzyme systems competing for a mutual substrate. University of Groningen (Thesis).

Kostoveckii YA & Zholdakova Z (1971) [On hygienic norm-setting in waterbodies.] Gig i Sanit, 7: 7-10 (in Russian).

Krug M, Ziegler H, & Straube G (1985) Degradation of phenolic compounds by the yeast *Candida tropicalis* HP-15: 1. Physiology and growth and substrate utilization. J Basic Microbiol, 25(2): 103-110.

Kumar V & Mukerjee D (1988) Phenol and sulfide induced changes in the ovary and liver of sexually maturing common carp, *Cyprinus carpio*. Aquat Toxicol, 13(1): 53-60.

Kuwata K, Uebori M, & Yamazaki Y (1980) Determination of phenol in polluted air as *p*-nitrobenzene azophenol derivative by reversed phase high performance liquid chromatography. Anal Chem, 52(6): 857-860.

Kwasniewska K & Kaiser KLE (1983) Toxicities of selected phenols to fermentative and oxidative yeasts. Bull Environ Contam Toxicol, 31(2): 188-194.

Laszczynska M, Barcew B, Wyczarska-Srak K, & Wozniak D (1983) Reversibility of histoenzymatic changes caused by phenol in the mouse kidney and liver. Folia Biol (Cracow), 31(1): 65-73.

Layiwola PJ & Linnecar DFC (1981) The biotransformation of ^{14}C phenol in some freshwater fish. Xenobiotica, 11(3): 167-171.

Leichnitz K ed. (1982) [Dräger test tubes phenol 5/a.] In: [Test tube pocket book; testing of air and technical analysis of gas using Dräger test tubes], 5th ed. Lubeck, Drägerwerk AG, p 114 (in German).

Leider M & Moser HS (1961) Toxicology of topical dermatology preparations. Arch Dermatol, 83: 928-929.

Leuenberger C, Ligocki MP, & Pankow JF (1985) Trace organic compounds in rain: 4. Identities, concentrations, and scavenging mechanisms for phenols in urban air and rain. Environ Sci Technol, 19(11): 1053-1058.

Lewin JF & Cleary WT (1982) An accidental death caused by the absorption of phenol through skin. A case report. Forensic Sci Int, 19(2): 177-179.

Lewis MA (1983) Effect of loading density on the acute toxicities of surfactants, copper and phenol to *Daphnia magna* Straus. Arch Environ Contam Toxicol, 12: 51-55.

Liao JTF (1980) Investigations of phenol tissue distribution, plasma protein binding, and liver subcellular fraction affinity. Diss Abstr Int, B41(5): 1736.

Liao JTF & Oehme FW (1980) Literature reviews of phenolic compounds. I. Phenol. Vet Hum Toxicol, 22: 160-164.

Liao JTF & Oehme FW (1981a) Tissues distribution and plasma protein binding of (^{14}C) phenol in rats. Toxicol Appl Pharmacol, 57(2): 220-225.

Liao JTF & Oehme FW (1981b) Plasma protein binding of phenol in dogs and rats as determined by equilibrium dialysis and ultrafiltration. Toxicol Appl Pharmacol., **57**(2): 226-230.

Lister J (1867; republished 1937) On a new method of treating compound fracture, abcess, etc., with observations on the condition of suppuration. Med Classics, **2**: 28-71.

Luten JB, Ritskes JM, & Weseman JM (1979) Determination of phenol, guaiacol and 4-methylguaiacol in wood smoke and smoked fish-products by gas-liquid chromatography. Z Lebensm.unters Forsch, **168**: 289-292.

McFarlane JC, Pfleeger T, & Fletcher J (1987) Transpiration effect on the uptake and distribution of bromacil, nitrobenzene and phenol in soybean plants. J Environ Qual, **16**(4): 372-376.

McGregor DB, Brown A, Cattanach P, Edwards I, McBride D, Riach C, & Caspary WJ (1988) Responses of the L5178Y tk+/tk- mouse lymphoma cell forward mutation assay: III. 72 Coded chemicals. Environ Mol Mutagen, **12**: 85-154.

Malcolm AR, Mills LJ, & Trosko JE (1985) Effects of ethanol, phenol, formaldehyde, and selected metabolites on metabolic cooperation between Chinese hamster V79 lung fibroblasts. Carcinogenesis: A compr survey, **8**: 305-318.

Mayes ME, Alexander HC, & Dill DC (1982) A study to assess the influence of age on the response of Fathead minnows in static acute toxicity tests. Bull Environ Contam Toxicol, **31**(2) 139-147.

Merliss RR (1972) Phenol marasmus. J Occup Med, **14**: 55-56.

Miksch K & Schürmann B (1988) [Toxicity assessment of zinc sulphate, copper sulphate and phenol by different methods.] Z Wasser Abwasser Forsch, **21**: 193-198 (in German with English abstract).

Miller DL & Ogilvie DM (1975) Temperature selection in brook trout (*Salvelinus fontinalis*) following exposure to DDT, PCB or phenol. Bull Environ Contam Toxicol, **14**: 545-551.

Minor JK & Becker BA (1971) A comparison of the teratogenic properties of sodium salycate, sodium benzoate and phenol. Abstracts of the 10th Annual Meeting of the Society of Toxicology No. 40. Toxicol Appl Pharmacol, **19**: 373.

Mitchell JR (1972) Mechanism of benzene-induced aplastic anaemia. Fed Proc, **30**: 561.

Model A (1889) Poisoning with concentrated phenol in a child suffering from diphtheria. Ther Mon.heft, **3**: 483-485.

Mogilnicka EM & Piotrowski JK (1974) [The exposure test for phenol in the light of field study.] Med Prac, **25**: 137-141 (in Hungarian).

Morimoto K & Wolff S (1980) Increase of sister chromatid exchanges and perturbations of cell division kinetics in human lymphocytes by benzene metabolites. Cancer Res, **40**: 1189-1193.

Morimoto K, Wolff S, & Koizumi A (1983) Induction of sister chromatid exchanges in human lymphocytes by microsomal activation of benzene metabolites. Mutat Res, 119(3-4): 355-360.

Morrison JE, Matthews D, Washington R, Fennessey PV, & Harrison M (1991) Phenol motor point blocks in children: Plasma concentrations and cardiac dysrhythmias. Anesthetiology, 75: 359-362.

Mukhitov B (1964) The effect of low phenol concentrations on the organism of man or animals and their hygienic evaluation. In: Levine BS ed. USSR literature on air pollution and related occupational diseases. Springfield, Virginia, US Department of Commerce, National Technical Information Service, pp 185-199 (NTIS/PB 64-11574).

Murphy JC, Osterberg RE, Seabaugh UM, & Bierbower GW (1982) Ocular irritancy responses to various pHs of acids and bases with and without irrigation. Toxicology, 23(4): 281-291.

Müting D, Keller HE, & Kraus W (1970) Quantitative colorimetric determination of free phenols in serum and urine of healthy adults using modified diaro-reactions. Clin Chem Acta, 27: 177-180.

Nagel R (1983) Species differences, influence of dose and application on biotransformation of phenol in fish. Xenobiotica, 13(2): 101-106.

Nagel R & Urich K (1983) Quinol sulfate, a new conjugate of phenol in goldfish. Xenobiotica, 13(2): 97-100.

Nagel R, Adler HJ, & Rao TK (1982) Induction of filamentation by mutagens and carcinogens in a lon mutant of *Escherichia coli*. Mutat Res, 105: 309-312.

Namkoong W, Loehr RC, & Malina JF Jr (1989) Effects of mixture and acclimation on removal of phenolic compounds in soil. J Water Pollut Control Fed, 61(2): 242-250.

Narotsky MG & Kavlock RJ (1993) A multidisciplinary approach to toxicological screening. II. Development toxicity. Research Triangle Park, North Carolina, US Environmental Protection Agency, Health Effects Research Laboratory (Report No. MS 91-237).

Nathan PW (1959) Intrathecal phenol to relieve spasticity in paraplegia. Lancet, 2: 1099.

NCI (1980) Bioassay of phenol for possible carcinogenicity. Bethesda, Maryland, US Department of Health Services, National Cancer Institute (Technical Report Series No. NCI-CG-TR-203).

Neuhauser EF, Durking PR, Malecki MR, & Anatra M (1986) Comparative toxicity of ten organic chemicals to four earthworm species. Comp Biochem Physiol, C83: 197-200.

Nicola RM, Branchflower R, & Pierce D (1987) Chemical contaminants in bottonfish. J Environ Health, 49: 342-347.

Niessen R, Lenoir D, & Boule P (1988) Phototransformation of phenol induced by excitation of nitrate ions. Chemosphere, 17(19): 1977-1984.

NIOSH (1984) NIOSH manual of analytical methods, 3rd ed. Cincinnati, Ohio, National Institute for Occupational Safety and Health.

NIOSH (1985a) NIOSH pocket guide to chemical hazards. Cincinnati, Ohio, National Institute for Occupational Safety and Health.

NIOSH (1985b) Method 8305. NIOSH manual of analytical methods, 3rd ed. Cincinnati, Ohio, Institute for Occupational Safety and Health.

Oehme FW (1969) A comparative study of the biotransformation and excretion of phenol. A doctoral dissertation. University of Missouri.

Oehme FW & Davis LE (1970) The comparative toxicity and biotransformation of phenol. Toxicol Appl Pharmacol, 17: 283.

Ogata M, Yamasaki Y, & Kawai T (1986) Significance of urinary phenyl sulfate and phenyl glucuronide as indices of exposure to phenol. Int Arch Occup Environ Health, 58: 197-202.

Oglesbi LA, Ebron-McCoy MT, Logson TR, Copeland F, Beyer PE, & Kavlock RJ (1992) *In vitro* embryotoxicity of a series of para-substituted phenols: structure, activity, and correlation with *in vivo* data. Teratology, 45: 11-33.

Ohtsuji H & Ikeda M (1972) Quantitative relationship between atmospheric phenol vapour and phenol in the urine of employees in Bakelite factories. Br J Ind Med, 29: 70-73.

Oksama M & Kristofferson R (1979) The toxicity of phenol to *Phoxinus phoxinus*, *Gammarus ducheni* and *Mesidota entomon* in brackish water. Ann Zool Fenn, 16(3): 209-216.

Olowska L, Oledzka-Slotwinska H, Lazczymska M, & Wozniak W (1980) Effect of phenol on hydrolytic and respiratory enzymes of the small intestinal mucosa of white mouse. Folia Morphol (Warsaw), 39(2): 149-158.

Oomens AC & Schuurhuis FG (1983) A method for the collection and determination of phenol and bisphenol A in air. Int Arch Occup Environ Health, 52(1): 43-48.

Oris JT, Winner RW, & Moore MV (1991) A four-day survival and reproduction toxicity test for *Ceriodaphnia dubia*. Environ Toxicol Chem, 10: 217-224.

Painter RB & Howard R (1982) The Hela DNA-synthesis inhibition test as a rapid screen for mutagenic carcinogens. Mutat Res, 92(1-2): 427-437.

Paradowski M, Anderzeiewski S, Zauisza B, & Wybrak-Wrobel T (1981) Excretion of phenol and *p*-cresol in urine as an element of evaluation of professional exposure to benzene and phenol in employees of refining and petrochemicals works in Plock. Przeglad Lekarski, 38(9): 663-666.

Paris DF, Wolfe NL, & Steen WC (1982) Structure-activity relationships in microbial transformation of phenols. Appl Environ Microbiol, 44: 153-158.

Parke DV & Williams RT (1953) Studies in detoxification. 54. The metabolism of benzene: (a) The formation of phenylglucuronide and phenylsulfuric acid from ^{14}C benzene. (b) The metabolism of ^{14}C phenol. Biochem J, 55: 337-340.

Parker JG (1984) The effects of selected chemicals and water quality on the marine polychaete *Ophryotrocha diadema*. Water Res, **18**(7): 865-868.

Pashin YV & Bakhitova LM (1982) Mutagenicity of benzo(a)pyrene and the antioxidant phenol at the HGPRT locus of V79 Chinese Hamster cells. Mutat Res, **104**(6): 389-393.

Patrick E, Maibach HI, & Burkhalter A (1985) Mechanisms of chemically induced skin irritation. I. Studies of time course, dose response, and components of inflammation in the laboratory mouse. Toxicol Appl Pharmacol, **81**: 476-490.

Pekari K, Vainotalo S, Heikkila P, Palotie A, Luotamo M, & Riihimaki V (1992) Biological monitoring of occupational exposure to low levels of benzene. Scand J Work Environ Health, **18**: 317-322.

Pellack-Walker P & Blumer JL (1986) DNA damage in L5178YS cells following exposure to benzene metabolites. Mol Pharmacol, **30**: 42-47.

Pellack-Walker P, Walker JK, Evans HH, & Blumer JL (1985) Relationship between the oxidation potential of benzene metabolites and their inhibitory effect on DNA synthesis in L5178YS cells. Mol Pharmacol, **28**: 560-566.

Petrov VI (1960) [Cases of phenol poisoning during coke slaking with phenol water.] Gig i Sanit, **25**: 60-62 (in Russian).

Phipps GL, Holcombe GW, & Fiandt JT (1981) Acute toxicity of phenol and substituted phenols to the fathead minnow. Bull Environ Contam Toxicol, **26**: 585-593.

Pickering GH & Henderson C (1966) Acute toxicity of some important petrochemicals to fish. J Water Pollut Control Fed, **38**: 1419-1429.

Pierce WM & Nerland DE (1988) Qualitative and quantitative analyses of phenol, phenylglucuronide, and phenylsulfate in urine and plasma by gas chromatography mass spectrometry. J Anal Toxicol, **12**: 344-347.

Piotrowski JK (1971) Evaluation of exposure to phenol, absorption of phenol vapour in the lungs and through the skin and excretion of phenol in urine. Br J Ind Med, **28**: 172-178.

Podolak GE, McKenzie RM, Rinehart DS, & Mazur JF (1981) A rapid technique for collection and analysis of phenol vapours. Am Ind Hyg Assoc J, **42**(10) 734-738.

Poirier MC, De Cicco BT, & Lieiberman MW (1975) Non-specific inhibition of DNA repair synthesis by tumour promotors on human diploid fibroblasts damaged with *N*-acetoxy-acetyl-aminofluorene. Cancer Res, **35**: 1392-1397.

Pool BL & Lin PZ (1982) Mutagenicity testing in the *Salmonella typhimurium* assay of phenolic compounds and phenolic fractions obtained from smoke-house condensates. Food Chem Toxicol, **22**: 383-391.

Potthast K (1976) Determination of phenols in smoked meat products. Advances in Smoking of Foods, Intern. Joint IUPAC/IUFoSt Symp., Warschau, pp. 39-44.

Potthast K (1982) Dark smoking at high smokehouse temperatures. Fleischwirtschaft, **62**: 1578-1582.

Powell GM, Miller JJ, Olavesen AH, & Curtis CG (1974) Liver as a major organ of phenol detoxification? Nature (Lond), **252**: 234-235.

Pratt RM & Willis WD (1985) *In vitro* screening assay for teratogens using growth inhibition of human embryonic cells. Proc Natl Acad Sci (USA), **82**: 5791-5794.

Price KS, Waggy GT, & Conway RA (1974) Brine shrimp bioassay and seawater BOD of petrochemicals. J Water Pollut Control Fed, **46**: 63-76.

Pullin TG, Pinkerton MN, Johnson RV, & Kilian DJ (1978) Decontamination of the skin of swine following phenol exposure: A comparison of the relative efficacy of water versus polyethylene glycol/industrial methylated spirits. Toxicol Appl Pharmacol, **43**: 199-206.

Quebbemann AJ & Anders MW (1973) Renal tubular conjugation and excretion of phenol and *p*-nitrophenol in the chicken: differing mechanism of renal transfer. J Pharmacol Exp Ther, **184**(3): 695-708.

Ramanathan M (1984) Water pollution. In: Mark HF, Othmer DF, Overberger CG, Seaborg GT, & Grayson M ed. Kirk-Othmer encyclopedia of chemical technology, 3rd ed. New York, Chichester Brisbane, Toronto, John Wiley & Sons, vol 24, p 299.

Ramli JB & Wheldrake JF (1981) Phenol conjugation in the desert hopping mouse, *Notomys alexis*. Comp Biochem Physiol, **C69**(2): 379-381.

Rao RSV & Nath KJ (1983) Biological effect of some poisons on Cantocamptus (Crustacea). Int J Environ Stud, **21**(3-4): 271-275.

Rao, PS, Durve US, Khangarot BS, & Srekhawat SS (1983) Acute toxicity of phenol, pentachlorophenol and sodium pentachlorophenate to a freshwater ostracod *Cypris subglubosa*. Acta Hydrochim Hydrobiol, **11**(4): 457-465.

Rapson WH, Nazar MA, & Butsky V (1980) Mutagenicity produced by aqueous chlorination of organic compounds. Bull Environ Contam Toxicol, **24**: 590-596.

Ravichandran S & Anantharaj B (1984) Effect of phenol on the phosphomonoesterases and ATPase activity in the fish *Sarotherodon mossambicus* (Peters) in saline waters. Proc Indian Acad Sci (Anim Sci), **93**(6): 557-563.

Razani H, Nanba K, & Murachi S (1986a) Acute toxic effect of phenol on zebra fish *Brachydanio rerio*. Bull Jpn Soc Sci Fish, **52**(9): 1547-1552.

Razani H, Nanba K, & Murachi S (1986b) Chronic toxic effect of phenol on zebra fish *Brachydanio rerio*. Bull Jpn Soc Sci Fish, **52**(9): 1553-1558.

Rea WJ, Pan Y, Laseter JL, Johnson AR, & Fenyves EJ (1987) Toxic volatile organic hydrocarbons in chemically sensitive patients. Clin Ecol, **5**(2): 70-74.

Reid GG, Ketterer PJ, Mawkinney H, & Glover P (1982) Exposure to phenol and endrin as a cause of skin ulcerations and nervous signs in pigs. Aust Vet J, **59**: 160.

Reynolds JH, Middlebrooks EJ, Porcella WB, & Grenney WJ (1978) Comparison of semi-continuous and continuous-flow bioassay. Progress in Water Technology, **9**(4): 897-909.

RIVM (1986) [Criteria Document: Phenol]. Bilthoven, The Netherlands, National Institute of Public Health and Environmental Protection (Document No. 738513002) (in Dutch; toxicology also available in English).

Roberts MS, Anderson RA, & Swarlrich J (1977) Permeability of human epidermis to phenolic compounds. J Pharm Pharmacol, 29(11): 677-683.

Roberts MS, Anderson RA, Swarlrich J, & Moore DE (1978) The percutaneous absorption of phenolic compounds: the mechanism of diffusion across the stratum corneum. J Pharm Pharmacol, 30: 486-490.

Rogers SCF, Burrows D, & Neill D (1978) Percutaneous absorption of phenol and methyl alcohol in magenta paint B.C.P. Br J Dermatol, 98: 559-560.

Rozich AF & Colvin RJ (1986) Effects of glucose on phenol biodegradation by heterogenous populations. Biotechnol Bioeng, 29(7): 965-971.

Rubin HE & Alexander M (1983) Effect of nutrients on the rates of mineralization of trace concentrations of phenol and p-nitrophenol. Environ Sci Technol, 17: 104-107.

Ruedemann R & Deichmann WB (1953) Blood phenol level after topical application of phenol containing preparations. J Am Med Assoc, 152: 506-509.

Ruessink RG & Smith LL (1975) The relationship of the 96 h LC_{50} of the lethal threshold concentration of hexavalent chromium, phenol and sodium pentachlorophenate for fathead minnows (*Pimephales promelas* Rafinesque). Trans Am Fish Soc, 3: 570.

Rusch HP, Bosch MS, & Boutwell RK (1955) The influence of irritants on mitotic activity and tumour formation in mouse epidermus. International Union against Cancer Newsletter, 11: 699-703.

Rushmore TH, Snyder R, & Kalf G (1984) Covalent binding of benzene and its metabolites to DNA in rabbit bone marrow mitochondria *in vitro*. Chem-Biol Interact, 49: 133-154.

Russell JW (1975) Analysis of air pollutants using sampling tubes and gas chromatography. Environ Sci Technol, 9(13): 1175-1178.

Ryser S & Ulmer G (1980) Etude de la pollution gazeuse par les résines synthétiques utilisées en fonderie. Fonderie, 35: 313-324.

Salamon MH & Glendenning OM (1957) Tumour promotion in mouse skin by sclerosing agents. Br J Cancer, 11: 434-444.

Sandage C (1961) Tolerance criteria for continuous inhalation exposure to toxic material. I. Effects on animals of 90-day exposure to phenol, CCl4, and a mixture of indole, skatole, H₂S, and methyl mercaptan. Dayton, Ohio, Wright-Patterson Air Force Base, US Air Force Systems Command, Aeronautical Systems Division (ASD Technical Report 61-519 (I); NTIS AD-268783).

Sanitary Epidemiological Station, Katowice (1991) Air pollution in the Katowice voivodeship in 1988-1990. Katowice, Poland, Regional Sanitary Epidemiological Station.

Sawahata T & Neal RA (1983) Biotransformation of phenol to hydroquinone and catechol by rat liver microsomes. Mol Pharmacol, 23(21): 453-460.

Schaumburg HH, Byck R, & Weller RO (1970) The effect of phenol on peripheral nerve. A histological and electrophysiological study. J Neuropathol Exp Neurol, 29: 615-630.

Schlicht MP, Moser VC, Sumrell BM, Berman E, & MacPhail RC (1992) Systemic and neurotoxic effects of acute and repeated phenol administration (Abstract No. 1047). Toxicologist, 12, 274.

Schmidt R & Maibach H (1981) Immediate and delayed onset "skip area" dermatitis presumed secondary to topical phenol exposure. Contact Dermatitis, 7(4): 199-202.

Schmieder PK & Henry TR (1988) Plasma binding of 1-butanol, phenol, nitrobenzene and pentachlorophenol in the rainbow trout and rat: a comparative study. Comp Biochem Physiol, C91(2): 413-418.

Schoenberg JB & Mitchell CA (1975) Airway disease caused by phenolic (phenol-formaldehyde) resin exposure. Arch Environ Health, 30: 574-577.

Schussler OF & Stern MA (1911) Gangrene of finger caused by five percent phenol ointment. J Am Med Assoc, 57: 628.

Schütz A & Wolfe D (1980) [Gases and vapours in foundries; measurement; evaluation; control.] Giesserei, 67: 68-73 (in German).

Seidel HJ, Barthel E, Schäfer F, Schad H, & Weber L (1991) Action of benzene metabolites on murine hematopoietic colony-forming cells *in vitro*. Toxicol Appl Pharmacol, 111, 128-131.

Selvakumar A & Hsieh HN (1988) Competitive adsorption of organic compounds by microbial biomass. J Environ Sci Health, A23(8): 729-744.

Schlicht MP, Moser VP, Sumrell BM, Berman E, & MacPhail RC (1992) Systematic and neurotoxic effects of acute and repeated phenol administration. Toxicologist, 12(1): 274.

Scott HD, Wolf DC, & Lavy TL (1982) Apparent absorption and microbial degradation of phenol by soil. J Environ Qual, 11: 107-112.

Shafer WE & Schönherr J (1985) Accumulation and transport of phenol, 2-nitrophenol, and 4-nitrophenol in plant cuticles. Ecotoxicol Environ Saf, 10(2): 239-252.

Shimp RJ & Pfaender FK (1987) Effect of adaptation to phenol on biodegradation of monosubstituted phenols by aquatic microbial communities. Appl Environ Microbiol, 53(7): 1496-1499.

Shimp RJ & Young RL (1987) Availability of organic chemicals for biodegradation in settled bottom sediments. Ecotoxicol Environ Saf, 15(1): 31-45.

Shirkey RJ, Kao J, Fry JR, & Bridges JW (1979) A comparison of xenobiotic metabolism in cells isolated from rat liver and small intestinal mucosa. Biochem Pharmacol, 28: 1461-1466.

Skare JA & Schrotel KR (1984) Alkaline elution of rat testicular DNA: Detection of DNA strand breaks after *in vivo* treatment with chemical mutagens. Mutat Res, **130**: 283-294.

Skvortsova NN & Vysochina IV (1976) Changes in biochemical and physiological indices in animals produced by the combined effect of benz[a]pyrene and phenol. Environ Health Perspect, **13**: 101-106.

Slabbert JL (1986) Improved bacterial growth test for rapid water toxicity screening. Bull Environ Contam Toxicol, **37**: 565-569.

Smart AC & Zannoni VC (1984) D.T. Diaphorase and peroxidase influence the covalent binding of the metabolites of phenol, the major metabolite of benzene. Mol Pharmacol, **26**: 105-111.

Smith MC (1984) Histological findings following intrathecal injections of phenol solutions for relief of pain. Br J Anaesth, **36**: 387-406.

Solbé JFLG, Cooper VA, Willis CA, & Mallett MJ (1985) Effects of pollutants in fresh waters on European non-salmoid fish. I: Non-metals. J Fish Biol, **27**(Suppl A): 197-207.

Solliman T (1957) A manual of pharmacology and its applications to therapeutics and toxicology, 8th ed. New York, Philadelphia, W.B. Saunders Company, pp 805-809.

Steele RH & Wilhelm DL (1966) The inflammatory reaction in chemical injury. I. Increased vascular permeability and erythema induced by various chemicals. Br J Exp Pathol, **47**: 612-623.

Southworth GR, Herbes SE, Franco PJ, & Giddings JM (1985) Persistence of phenols in aquatic microcosms receiving chronic inputs of coal-derived oil. Water Air Soil Pollut, **24**(3): 283-296.

Spicer CW, Riggin RM, Holdren MW, DeRoos FL, & Lee RN (1985) Atmospheric reaction products from hazardous air pollutant degradation. Research Triangle Park, North Carolina, US Environmental Protection Agency (EPA-600/3-85-028; NTIS PB85-185841/XAB).

Spoelstra SF (1978) Degradation of tyrosine in anaerobically stored piggery wastes and in pig feces. Appl Environ Microbiol, **36**: 631-638.

Stajduhar-Caric Z (1968) Acute phenol poisoning. J Forensic Med, **15**: 41-42.

Stephenson RR (1983) Effects of water hardness, water temperature and size of the test organism on the susceptibility of the freshwater shrimp *Gammarus pulex* (L.) to toxicants. Bull Environ Contam Toxicol, **31**(4): 459-466.

Sturtevant MJ (1952) Studies on the mutagenicity of phenol. Journal of Heredity, **43**: 217-219.

SRI (1982) Economics handbook. Stanford, Stanford Research Institute.

Subrahmanyam VV & O'Brien PJ (1985) Peroxidase-catalyzed binding of uniformly carbon-14-labeled phenol to DNA. Xenobiotica, **15**(10): 859-872.

Subrahmanyam V, Sadler A, Suba E, & Ross D (1989) Stimulation of *in vitro* bioactivation of hydroquinone by phenol in bone marrow cells. Drug Metab Dispos, 17(3): 348-349.

Subrahmanyam VV, Kolachana P, & Smith MT (1991) Hydroxylation of phenol to hydroquinone catalyzed by a human myeloperoxidase-superoxide complex: possible implications in benzene-induced myelotoxicity. Free Radic Res Commun, 15, 285-296.

Suzuki T & Kisara K (1985) Enhancement of phenol-induced tremor caused by central monoamine depletion. Pharmacol Biochem Behav, 22: 153-155.

Takemori AE & Glowacki GA (1962) Studies on glucuronide synthesis in rats chronically treated with morphine and phenol. Biochem Pharmacol, 11: 867-870.

Tetra Tech. Inc. (1986) Development of sediment quality values for Pudget Sound (unnamed report prepared for Pudget Sound Dredged Disposal Analysis and Pudget Sound Estuary Program). Bellevue, Washington, Tetra Tech. Inc.

Thompson ED & Gibson DP (1984) A method for determining the maximum tolerated dose for acute *in vivo* cytogenetic studies. Food Chem Toxicol, 22(8): 665-676.

Thornton-Manning JR, Jones DD, & Federle TW (1987) Effects of experimental manipulation of environmental factors on phenol mineralization in soil. Environ Toxicol Chem, 6: 615-621.

Tibbles BJ & Baecker AAW (1989) Effects and fate of phenol in simulated landfill sites. Microb Ecol, 17(2): 201-206.

TNO (1978) [Emissions during various processes in foundries.] Delft, The Netherlands, Dutch Organization for Applied Scientific Research (TNO) (Report No. M78/40/07536/Jul/SCI) (in Dutch).

Tonapi GT & Varghese G (1987) Cardio-physiological responses of some selected cladocerans to three common pollutants. Arch Hydrobiol, 110(1): 59-65.

Toth L (1982) [Transfer of phenols from tobacco smoke into meat products.] Fleischwirtschaft, 62: 1389-1402 (in German).

Tranvik L, Larsson P, Okla L, & Regnell O (1991) In situ mineralization of chlorinated phenols by pelagic bacteria in lakes of different humic content. Environ Toxicol Chem, 10: 195-200.

Truppman ES & Ellenby JD (1979) Major electrocardiograph changes during chemical face peeling. Plast Reconstr Surg, 63: 44-48.

Tschech A & Fuchs G (1987) Anaerobic degradation of phenol by pure cultures of newly isolated denitrifying pseudomonads. Arch Microbiol, 148(3): 213-217.

Tunek A, Olofsson J, & Berlin M (1981) Toxic effects of benzene and benzene metabolites on granulopoietic stem cells and bone marrow cellularity in mice. Toxicol Appl Pharmacol, 59(1): 149-156.

Turtle WRM & Dolan T (1922) A case of rapid and fatal absorption of carbolic acid through the skin. Lancet, 2: 1273-1274.

UBA (1981) [Air pollution control '81. Developments, state of the art, tendencies.] Berlin, Federal Office for the Environment (in German).

United Nations (1980) Yearbook or industrial statistics. Volume II: Commodity product data. New York, United Nations.

Ursin C (1985) Degradation of organic chemicals at trace levels in seawater and marine sediment. The effects of concentration of the initial fractional turnover rate. Chemosphere, 14(10): 1539-1550.

US EPA (1980) Phenol. Ambient water quality criteria. Washington, DC, US Environmental Protection Agency (EPA 440/5-80-066).

US EPA (1986a) Method 8010. Phenols. In: Test methods for evaluating solid waste - Physical/chemical methods, 3rd ed. Washington, DC, US Environmental Protection Agency, Office of Solid Waste and Emergency Response, pp 8040/1-8050/17 (EPA Report No. SW-846).

US EPA (1986b) Method 8250. Gas chromatography/mass spectrometry for semi-volatile organics: Packed column technique. In: Test methods for evaluating solid waste - Physical/chemical methods, 3rd ed. Washington, DC, US Environmental Protection Agency, Office of Solid Waste and Emergency Response, pp 8250/1-8250/32 (EPA Report No. SW-846).

US EPA (1986c) Method 8270. Gas chromatography/mass spectrometry for semi-volatile organics: Capillary column technique. In: Test methods for evaluating solid waste - Physical/chemical methods, 3rd ed. Washington, DC, US Environmental Protection Agency, Office of Solid Waste and Emergency Response, pp 8270/1-8270/34 (EPA Report No. SW-846).

US EPA (1992) Phenol - Drinking water health advisory. Washington, DC, US Environmental Protection Agency, Office of Water.

Van Duuren BL & Goldschmidt BM (1976) Carcinogenic and tumour-promoting agents in tobacco carcinogenesis. J Natl Cancer Inst, 56: 1237-1242.

Van Duuren BL, Sivah A, Langretti J, & Goldschmidt BM (1968) Initiators and promotors in tobacco carcinogenesis. Natl Cancer Inst Monogr, 28: 173-180.

Van Duuren BL, Blazej T, Goldschmidt BM, Katz C, Melchionne S, & Sivah A (1971) Carcinogenesis studies on mouse skin and inhibition of tumour induction. J Natl Cancer Inst, 46: 1039-1044.

Van Duuren BL, Katz C, & Goldschmidt BM (1973) Cocarcinogenesis agents in tobacco carcinogenesis. J Natl Cancer Inst, 51: 703-705.

Van Gemert LJ (1984) Compilation of odour threshold values in air: Supplement V. Zeist, The Netherlands, CIVO-TNO (Report No. 84.220).

Van Gemert LJ & Van Nettenbreijer (1977) Compilation of odour threshold values in air and water. Zeist, The Netherlands, CIVO/RID.

Veith GD & Kosian P (1983) Estimating bioconcentration potential from octanol/water partition coefficients. In: Physical behaviour of PCB's in the Great Lakes. Ann. Arbor, Michigan, Ann Arbor Science Publishers, Inc., pp 269-282.

Verma SR, Rani S, Tyagi AK, & Dalela RC (1980) Evaluation of acute toxicity of phenol and its chloro- and nitro-derivates to certain teleosts. Water Pollut, 14: 95-102.

Vernot EH, MacEwen JD, Haun CC, & Kinkead ER (1977) Acute toxicity and skin corrosion data for some organic and inorganic compounds and aqueous solutions. Toxicol Appl Pharmacol, 42: 417-423.

Veronesi B, Padilla S, & Newland D (1986) Biochemical and neuropathological assessment of triphenyl phosphite in rats. Toxicol Appl Pharmacol, 83: 203-210.

Verschueren K (1983) Handbook of environmental data on organic chemicals, 2nd ed. New York, Van Nostrand Reinhold Company.

Volskay VT Jr & Grady CPL (1988) Toxicity of selected RCRA compounds to activated sludge microorganisms. J Water Pollut Control Fed,. 60(10): 1850-1856.

Von Oettingen WF (1949) Phenol and its derivatives. The relation between their chemical constitution and their effect on the organisms. Washington, DC, National Institutes of Health Bulletin.

Von Oettingen WF & Sharples NE (1946) The toxicity and toxic manifestation of 2,2 bis-(p-chlorophenyl)-1,1,1-trichloroethane (DDT) as influenced by chemical changes in the molecule. J Pharmacol Exp Ther, 88: 400-413.

Wajon JE, Rosenblatt DH, & Burrows EP (1982) Oxidation of phenol and hydroquinone by chlorine dioxide. Environ Sci Technol, 16: 396-402.

Wang WC (1986) Comparative toxicology of phenolic compounds using root elongation method. Environ Toxicol Chem, 5(10): 891-896.

Wang YT, Suidan MT, Pfeiffer JT, & Najm I (1988) Effects of some alkyl phenols on methanogenic degradation of phenol. Appl Environ Microbiol, 54(5): 1277-1279.

Wangenheim J & Bolcsfoldi G (1988) Mouse lymphoma L5178Y thymidine kinase locus assay of 50 compounds. Mutagenesis, 3(3): 193-205.

Warner MA & Harper JV (1985) Cardiac dysrhythmias associated with chemical peeling with phenol. Anesthesiology, 62: 366-367.

Weast RC, ed. (1987) Handbook of chemistry and physics, 68th ed. Boca Raton, Florida, CRC Press, Inc., p C-406.

Weitering JG, Krijgsheld KR, & Mulder GJ (1979) The availability of inorganic sulfate as a rate limiting factor in the sulfate conjugation or xenobiotics in the rat? Biochem Pharmacol, 28: 757-762.

Wheldrake JF, Baudinette RV, & Hewitt S (1978) The metabolism of phenol in a desert rodent Notomys alexis. Comp Biochem Physiol, C61: 103-107.

Wiggins BA & Alexander M (1988) Role of chemical concentration and second carbon sources in acclimation of microbial communities for biodegradation. Appl Environ Microbiol, **54**(11): 2803-2807.

Wilcosky TC & Tyroler HA (1983) Mortality from heart disease among workers exposed to solvents. J Occup Med, **25**: 879-885.

Wilcosky TC, Checkoway H, Marshall EG, & Tyroler HA (1984) Cancer mortality and solvent exposures in the rubber industry. Am Ind Hyg Assoc J, **45**: 809-811.

Wild D, Eckhardt K, Gocke E, & King MT (1980) Comparative results on short-term *in vitro* and *in vivo* mutagenicity tests obtained with selected environmental chemicals. Berlin, Heidelberg, New York, Springer Verlag, pp 170-178.

Williams RT (1938) Studies in detoxification. I. The influence of (a) dose and (b) o, m, p, substitution on the sulfate detoxification of phenol in the rabbit. Biochem J, **32**: 878-887.

Williams RT (1959) Detoxication mechanisms, 2nd ed. New York, Chichester, Brisbane, Toronto, John Wiley and Sons, pp 237-295.

Windus-Podehl G, Lyftog C, Zieve L, & Brunner G (1983) Encephalopathic effect in rats. J Lab Clin Med, **101**(4): 586-592.

Winkler HD (1981) [Methods for reduction of emission of air pollutants in production of chipboards.] Gesundheits-Ing, **102**: 295-299 (in German).

Wood KA (1978) The use of phenol as a neurolytic agent: a review. Pain, **5**: 205-225.

Woodruff RC, Mason JM, Valencia R, & Zimmering S (1985) Chemical mutagenesis testing in Drosophila. V. Results of 53 coded compounds tested for the National Toxicology Program. Environ Mutagen, **7**: 677-702.

Wynder EL & Hoffmann D (1961) A study of tobacco carcinogenesis - the role of the acid fractions as promoters. Cancer, **14**: 1306-1315.

Wysowski DK, Flynt JW, Goldfield M, Altman R, & Davis AT (1978) Epidemic neonatal hyperbilirubinemia and use of a phenolic disinfectant detergent. Pediatrics, **61**(2): 165-170.

Young LY & Rivera MD (1985) Methanogenic degradation of 4 phenolic compounds. Water Res, **19**(10): 1325-1332.

Zavorovskaya NA & Nekhorosheva EV (1981) [Spectrometric determination of phenol and aniline in air and solutions simulating biological media.] Zh Anal Khim, **36**(9): 1808-1812 (in Russian).

RESUME

1. Identité, propriétés physiques et chimiques et méthodes d'analyse

Le phénol se présente sous la forme d'un solide cristallin blanc qui fond à 43 °C et se liquéfie par contact avec l'eau. Il dégage une odeur âcre caractéristique et possède une saveur forte et piquante. Il est soluble dans la plupart des solvants organiques; sa solubilité dans l'eau est limitée à la température ambiante; au-dessus de 68 °C il est entièrement soluble dans l'eau. Le phénol est modérément volatil à la température ambiante. C'est un acide faible, et, sous sa forme ionisée, il est très sensible aux réactions de substitution électrophile et à l'oxydation.

On peut recueillir le phénol dans des échantillons prélevés dans l'environnement par absorption dans une solution de soude ou par adsorption sur un solide. La désorption s'effectue par acidification, entraînement à la vapeur et extraction à l'éther (à partir des solutions) ou par voie thermique ou en phase liquide (lorsqu'il est adsorbé sur un solide). Les méthodes d'analyse les plus importantes sont la chromatographie en phase gazeuse avec détection par ionisation de flamme ou capture d'électrons ou encore la chromatographie liquide à haute performance avec détection en lumière ultra-violette. La limite de détection la plus basse qui ait été signalée dans l'air est de 0,1 μg par m^3. On peut doser le phénol dans le sang et les urines; dans les échantillons d'urine, on a fait état d'une limite de détection de 0.5 μg/litre.

2. Sources d'exposition humaine et environnementale

Le phénol est un constituant du goudron de houille et il se forme au cours de la décomposition naturelle des substances organiques. La majeure partie du phénol présent dans l'environnement provient cependant de l'activité humaine. Les sources potentielles en sont la production et l'utilisation, tel quel ou sous forme de dérivés, en particulier de résines phénoliques et de caprolactame, les gaz d'échappement, la combustion de bois de construction et la fumée de cigarette. Une autre source potentielle est constituée par la dégradation atmosphérique du benzène sous l'action de la lumière, la présence de phénol dans le purin pouvant également fortement contribuer à sa concentration dans l'atmosphère. Les dérivés du benzène et du phénol peuvent, par conversion *in vivo*, constituer une source d'exposition humaine endogène.

La production de phénol dans le monde s'est montrée relativement constante pendant les années 1980, les Etats-Unis d'Amérique en étant le premier producteur. Il est principalement utilisé pour la fabrication des résines phénoliques, du bisphénol A et de la caprolactame. On en connaît également un certain nombre d'applications médicales et pharmaceutiques.

3. Transport, distribution et transformation dans l'environnement

Les principales émissions de phénol se produisent dans l'air. La majeure partie du phénol présent dans l'atmosphère finit par être dégradée par voie photochimique en dihydroxybenzènes, nitrophénols et dérivés résultant de l'ouverture du cycle, avec une demi-vie estimative de 4 à 5 heures. Une petite quantité est éliminée de l'air par dépôt humide (pluie). Le phénol devrait présenter une forte mobilité dans le sol mais son transport et sa réactivité peuvent être affectés par le pH.

Le phénol présent dans l'eau et le sol peut être décomposé par des réactions abiotiques ainsi que par l'activité microbienne en un certain nombre de produits, dont les plus importants sont le dioxyde de carbone et le méthane. La part des réactions biologiques dans la décomposition globale du phénol dépend de nombreux facteurs tels que la concentration, l'acclimatation, la température et la présence d'autres composés.

4. Concentrations dans l'environnement et exposition humaine

On ne possède aucune donnée sur la concentration atmosphérique du phénol. La concentration de fond est vraisemblablement inférieure à 1 ng/m³. Les valeurs en milieu urbain et suburbain varient de 0,1 à 8 μg/m³ alors que dans les zones où prédominent les sources de phénol (zones industrielles) les concentrations signalées peuvent être jusqu'à 100 fois plus élevées. On a décelé du phénol dans l'eau de pluie, les eaux superficielles et les eaux souterraines mais les données sont très rares. On a fait état de concentrations élevées de phénol dans des sédiments et des eaux souterraines par suite de pollution industrielle.

Il peut y avoir exposition professionnelle au phénol lors de la production de ce produit et de ses dérivés, lors de l'enduction avec des résines phénoliques (industrie du bois et métallurgie) et lors d'un certain nombre d'autres activités industrielles. La concentration la plus élevée qui ait été signalée (jusqu'à 88 mg/m³) concernait des ouvriers de l'ex-URSS employés à l'extinction du coke avec des eaux usées contenant du phénol. La plupart des autres concentrations évoquées ne dépassaient 19 mg/m³.

En ce qui concerne la population dans son ensemble, c'est la fumée de cigarette et les aliments fumés qui constituent les plus importantes sources d'exposition au phénol, si l'on excepte l'exposition par voie atmosphérique. L'exposition par l'eau de boisson ou la consommation par inadvertance de produits alimentaires contaminés devraient rester faibles; le phénol a en effet une odeur et une saveur désagréables, ce qui dans la plupart des cas devrait alarmer le consommateur.

5. Cinétique et métabolisme

Le phénol est facilement absorbé par toutes les voies d'exposition. Après absorption, il se répartit rapidement dans l'ensemble des tissus.

Une fois résorbé, il forme essentiellement des conjugués avec l'acide glucuronique et l'acide sulfurique, et, dans une moindre mesure, des hydroxylates avec le catéchol et l'hydroquinone. Il y a également conjugaison avec les phosphates. La formation de métabolites réactifs (4,4-biphénol et diphénoquinone) a été mise en évidence lors d'études *in vitro* portant sur des neutrophiles et des leucocytes humains activés.

La proportion relative de glucuronides et de sulfo-conjugués varie avec la dose et l'espèce animale. Chez le rat, on a observé qu'en augmentant la dose de phénol, la formation de sulfo-conjugués l'emportait sur celle de glucuro-conjugués.

C'est essentiellement dans le foie, les poumons et au niveau de la muqueuse gastro-intestinale que le phénol est métabolisé. Le rôle relatif joué par ces divers tissus dépend de la voie d'administration et de la dose.

Des études *in vivo* et *in vitro* ont montré que le phénol se fixait aux protéines tissulaires et plasmatiques par liaison covalente. Certains métabolites du phénol se lient également aux protéines.

C'est principalement par la voie urinaire que la phénol est excrété chez l'animal et l'homme. Le taux d'excrétion urinaire varie selon la dose, la voie d'administration et l'espèce. Une faible proportion est excrétée dans les matières fécales et l'air expiré.

6. Effets sur les animaux d'expérience et les systèmes d'épreuve *in vitro*

Le phénol présente une toxicité aiguë modérée pour les mammifères. La DL_{50} par voie orale varie chez les rongeurs de 300 à 600 mg de phénol/kg de poids corporel. La DL_{50} dermique varie respectivement de 670 à 1400 mg/kg de poids corporel chez le rat et le lapin et pour le rat, la CL_{50} à 8h. par voie respiratoire est supérieure à 900 mg de phénol/m^3. Après exposition aiguë, les symptômes cliniques sont une hyperexcitabilité neuromusculaire, des convulsions graves, une nécrose de la peau et des muqueuses de la gorge et l'on note également des effets au niveau des poumons, des fibres nerveuses, des reins, du foie et de la pupille (réflexe photomoteur).

Les solutions de phénol sont agressives pour la peau et les yeux. A l'état de vapeur, le phénol peut irriter les voies respiratoires. On est fondé à croire que le phénol n'agit pas comme sensibilisateur cutané.

Les effets les plus importants relevés lors d'études à court terme sur l'animal consistaient en neurotoxicité, lésions hépatiques et rénales, troubles respiratoires et retard de croissance. A des doses orales quotidiennes de 40 mg/kg ou davantage on a observé des effets néphrotoxiques chez le rat. Chez la même espèce, il y avait une évidente hépatotoxicité aux doses supérieures ou égales à 100 mg/kg/jour. Lors d'une étude limitée de 14 jours sur des rats, on a obtenu, pour la dose par voie orale sans effets nocifs observables, une valeur de 12 mg/kg/jour, le critère retenu étant les effets sur le rein. Dans cette expérience, il y avait encore inhibition du myosis (réaction de l'iris à un stimulus lumineux) à la dose quotidienne de 4 mg/kg; toutefois, l'importance médicale de cette observation demeure incertaine. On a signalé la présence de certaines altérations biologiques au niveau de la muqueuse intestinale et des reins chez des souris recevant des doses

quotidiennes inférieures à 1 mg/kg, observation dont l'importance toxicologique n'est pas non plus bien claire.

Il n'y a pas eu d'études satisfaisantes sur la toxicité du phénol pour la fonction de reproduction. Toutefois, la toxicité du phénol paraît se manifester par son action délétère sur le développement du rat et de la souris. Lors de deux études au cours desquelles des rats ont reçu des doses multiples de phénol, on a obtenu, pour la dose sans effets nocifs observables, une valeur de 40 mg/kg/jour (pour la dose la plus faible sans effets nocifs observables, cette valeur était de 53 mg/kg/jour) et de 60 mg/kg/jour respectivement (dans ce cas, la dose la plus faible sans effets nocifs observables était de 120 mg/kg/jour). Chez la souris, la dose sans effets nocifs observables était de 140 mg/kg/jour (dose minimale sans effets nocifs observables: 280 mg/kg/jour).

La plupart des tests de mutagénicité bactérienne ont donné des résultats négatifs. Cependant, des épreuves effectuées *in vitro* sur des cellules mammaliennes ont révélé la présence de mutations, de lésions chromosomiques et d'effets sur l'ADN. Le phénol est sans effet sur la communication intercellulaire (mesurée par la coopération métabolique) dans des cultures de cellules mammaliennes. Un certain nombre d'études ont mis en évidence l'induction de micro-noyaux dans des cellules de moelle osseuse murine. Toutefois, les études sur la souris n'ont pas révélé la présence de micro-noyaux à doses plus faibles.

Deux études de cancérogénicité ont été effectuées sur des rats et des souris mâles et femelles à qui l'on administrait du phénol mêlé à leur eau de boisson. On n'a observé d'affections malignes (à savoir cancers médullaires de la thyroïde, leucémies) que chez les rats mâles soumis à de faibles doses. On n'a pas effectué d'études de cancérogénicité en bonne et due forme utilisant la voie percutanée ou la voie respiratoire. Des études de cancérogénicité en deux phases ont montré que le phénol pouvait se comporter comme un agent tumoro-promoteur lorsqu'on l'appliquait à plusieurs reprises sur la peau de la souris.

7. Effets sur l'homme

Des cas bien documentés d'exposition humaine au phénol par la voie percutanée, buccale ou intraveineuse, ont donné lieu à l'observation d'effets indésirables très divers. Il a été fait état d'une irritation des voies gastro-intestinales après ingestion de phénol. Après exposition de la peau, les effets observés localement vont d'un blémissement cutané indolore ou d'un érythème à la corrosion et à la nécrose profonde. Parmi les effets généraux, on a noté les troubles suivants: arythmies cardiaques, acidose métabolique, hyperventilation, détresse respiratoire, insuffisance rénale aiguë, lésions rénales, urines foncées, méthémoglobinémie, troubles neurologiques (notamment des convulsions), choc cardio-vasculaire, coma et mort. La dose orale la plus faible qui ait entraîné un décès humain était de 4,8 g; la mort est survenue dans les 10 minutes.

Le risque d'intoxication par inhalation de vapeurs de phénol est connu depuis longtemps, mais on n'a pas signalé de décès consécutif à ce type d'accident. Les symptômes produits par l'inhalation de phénol consistent notamment en anorexie, perte de poids, maux de tête, vertiges, salivation et coloration foncée des urines.

Le phénol n'est pas un agent sensibilisateur.

Le seuil olfactif pour l'homme serait 0,021 à 20 mg/m^3 d'air. Pour le phénol en solution aqueuse, on a fait état d'un seuil olfactif de 9 mg/litre, le seuil gustatif étant de 0,3 mg/litre d'eau.

On ne dispose pas de données suffisantes sur le pouvoir cancérogène du phénol.

8. Effets sur les êtres vivants dans leur milieu naturel

Lors d'études portant sur une seule espèce de bactéries, on a obtenu, pour la CE_{50} relative à l'inhibition de la croissance, des valeurs allant de 244 à 1600 mg de phénol/litre. On a constaté que le seuil de toxicité se situait à 64 mg de phénol/litre. Pour les protozoaires et les champignons, les valeurs étaient du même ordre que pour les bactéries; pour les algues elles étaient un peu inférieures. Le phénol est toxique pour les organismes dulçaquicoles supérieurs. Pour les crustacés et les poissons, les valeurs les plus faibles de la CL_{50} ou de CE_{50} se situent entre 3 et 7 mg de phénol/litre. Les données concernant la toxicité aiguë du phénol pour les organismes marins sont comparables à celles dont on dispose au sujet des organismes d'eau douce. Des études à long terme sur des crustacés et diverses espèces de poissons ont révélé des différences de sensibilité remarquables; c'est ainsi que les valeurs de la CL_1 provenant d'épreuves sur des embryons et des larves de *Salmo* et de *Carassius* se sont révélées très inférieures (respectivement 0,2 et 2 μg de phénol/litre) aux valeurs correspondantes pour d'autres espèces de poissons (concentration sans effet létal observable, 2,2-6,1 mg/litre) et d'amphibiens, ou tirées d'études sur la reproduction des crustacés (concentration sans effet létal observable, 10 mg de phénol/litre). On ne dispose pas de données sur des épreuves à long terme qui auraient été pratiquées sur des organismes marins.

En général le facteur de bioconcentration du phénol chez les divers types d'organismes aquatiques est très bas (<1-10) encore qu'on ait signalé parfois des valeurs plus élevées (jusqu'à 2200). Il est donc vraisemblable que le phénol ne subit pas d'accumulation biologique importante.

Les données dont on dispose au sujet de la destinée et des effets du phénol chez les organismes terrestres sont très peu nombreuses. La CE_{50} à 120 h. est de 120 à 170 mg/litre pour le millet et lors d'une épreuve par contact, on a obtenu pour la CL_{50} chez le lombric, une valeur comprise entre 2,4 et 10,6 μg/cm³.

9. Résumé de l'évaluation

9.1 Santé humaine

La population dans son ensemble est essentiellement exposée au phénol par la voie respiratoire. Par voie orale, il peut y avoir exposition répétée par suite de la consommation d'eau de boisson contaminée ou d'aliments fumés.

On ne dispose pas de données suffisantes pour déterminer l'ampleur de l'exposition de la population générale, mais on peut donner une limite estimative supérieure de l'absorption quotidienne. En mettant les choses au pire, on peut considérer que l'exposition maximale se produit chez un individu qui inhale en permanence de l'air fortement contaminé et consomme souvent des aliments fumés ou de l'eau de boisson qui contient du phénol à des concentrations atteignant le seuil gustatif. On a calculé que la dose quotidienne ingérée maximale totale estimative pour un individu de ce genre pesant 70 kg était de 0,1 mg/kg de poids corporel.

Les valeurs de la dose sans effets nocifs observables obtenues par expérimentation animale en prenant comme critères les troubles rénaux et les effets sur le développement étaient, chez le rat, de l'ordre de 12 à 40 mg/kg de poids corporel et par jour. En utilisant un coefficient d'incertitude de 200, on peut recommander comme limite supérieure de la dose journalière totale une valeur située entre 60 et 200 µg/kg de poids corporel. En prenant pour l'homme une dose quotidienne limite de 100 µg/kg de poids corporel, on peut conclure que l'exposition de toutes origines au phénol de la population dans son ensemble se situe en-dessous de ces valeurs.

On peut être préoccupé par le fait que, selon certaines données, le phénol pourrait être génotoxique et que d'autre part, on ne possède pas suffisamment de résultats pour écarter avec certitude l'éventualité que le phénol soit cancérogène. L'évaluation de ce composé doit être revue périodiquement.

9.2 Environnement

Le phénol ne subit probablement pas d'accumulation biologique importante. Il est toxique pour les organismes aquatiques; en appliquant la méthode modifiée de l'Agence de protection de l'environnement des Etats-Unis, on peut considérer que la concentration préoccupante de cette substance dans l'environnement est de 0,02 µg/litre. On manque de données suffisantes sur son action chez les plantes et les organismes terrestres.

Il peut y avoir transport du phénol d'un compartiment à l'autre de l'environnement par dépôt humide ou par lessivage du sol. En général, ce composé ne devrait pas persister dans l'environnement. Les rares données dont on dispose sur l'exposition ne permettent pas d'évaluer le risque qu'il présente pour les écosystèmes aquatiques ou terrestres. Toutefois, en tenant compte de la valeur de sa concentration préoccupante pour l'environnement aquatique, il est raisonnable de considérer qu'en cas de contamination par le phénol des eaux de surface ou des eaux marines, il y a un risque pour les organismes aquatiques.

RESUMEN

1. Identidad, propiedades físicas y químicas, métodos analíticos

El fenol es un sólido cristalino, blanco, que funde a 43 °C y se licúa al contacto con el agua. Posee un olor acre característico y un sabor ardiente fuerte. Es soluble en la mayor parte de los disolventes orgánicos. A temperatura ambiente, su solubilidad en agua es limitada; por encima de 68 °C es completamente hidrosoluble. El fenol es moderadamente volátil a temperatura ambiente. Es un ácido débil y, en su forma ionizada, muy sensible a las reacciones de sustitución electrofílica y a la oxidación.

El fenol se puede obtener a partir de muestras ambientales por absorción en una solución de NaOH o en contacto con sorbentes sólidos. La desorción se lleva a cabo por acidificación, destilación al vapor y extracción con éter (a partir de soluciones) o mediante desorción térmica o líquida (a partir de sorbentes sólidos). Las técnicas analíticas más importantes son la cromatografía de gases en combinación con la detección de ionización por conductor y de captura de electrones, y la cromatografía en fase líquida, de alta presión, en combinación con la detección por luz ultravioleta. En el aire, el límite de detección más bajo que se haya notificado es de 0,1 μg/m^3. Se puede determinar la presencia de fenol en la sangre y la orina; en muestras de orina se ha registrado un límite de detección de 0,5 μg/litro.

2. Fuentes de exposición humana y ambiental

El fenol es uno de los componentes del alquitrán de hulla y se forma durante la descomposición natural de materiales orgánicos. No obstante, la mayor parte del fenol presente en el medio ambiente es de origen antropogénico. Algunas fuentes potenciales son la producción y el uso de fenol y de sus productos, especialmente plásticos fenólicos y caprolactama, los gases de escape, la quema de leña y el humo de los cigarrillos. Otra fuente potencial es la degradación atmosférica del benceno por la influencia de la luz, si bien la presencia del fenol en los purines puede asimismo tener considerable influencia en sus niveles atmosféricos. Los derivados del benceno y del fenol pueden, mediante una conversión *in vivo*, constituir una fuente de exposición humana endógena a fenol.

Según parece, la producción mundial de fenol fue bastante regular a lo largo del decenio de 1980, en que los Estados Unidos fueron el productor más importante. Se usa principalmente como materia básica de las resinas fenólicas, del bisfenol A y de la caprolactama. También se le conocen algunas aplicaciones médicas y farmacéuticas.

3. Transporte, distribución y transformación en el medio ambiente

Las principales emisiones de fenol van al aire. La mayor parte del fenol existente en la atmósfera se degradará mediante reacciones fotoquímicas frente a los dihidroxibencenos, los nitrofenoles y los productos de rotura del anillo, con una semivida, estimada en 4 a 5 hs. Una parte menor desaparecerá del aire por deposición hídrica (lluvia). Se piensa que el fenol es móvil en el suelo, pero el pH puede influir en el transporte y la reactividad.

El fenol presente en el agua y el suelo puede degradarse por reacciones abióticas, así como por la actividad microbiana, dando lugar a un número de compuestos, los más importantes de los cuales son el dióxido de carbono y el metano. La proporción entre la biodegradación y la degradación general del fenol está determinada por múltiples factores, como la concentración, la aclimación, la temperatura y la presencia de otros compuestos.

4. Niveles ambientales y exposición humana

No se dispone de datos sobre los niveles atmosféricos de fenol. Se supone que los niveles básicos son inferiores a 1 ng/m³. Los niveles urbanos y suburbanos oscilan entre 0,1 y 8 µg/m³, mientras que se ha notificado que las concentraciones en las zonas próximas al foco de emisión (industria) alcanzan magnitudes cien veces superiores. Se ha detectado fenol en la lluvia y en las aguas superficiales y subterráneas, pero los datos son muy escasos. En sedimentos y aguas subterráneas se han notificado niveles elevados de fenol debidos a la contaminación industrial.

La exposición profesional al fenol puede tener lugar durante la producción del mismo y de sus derivados, la aplicación de resinas fenólicas (industrias maderera y siderúrgica) y algunas otras actividades industriales. La concentración más alta (hasta 88 mg/m³) se ha notificado en relación con trabajadores de la antigua Unión Soviética que apagaban el coque con aguas residuales que contenían fenol. La mayor parte de las restantes concentraciones notificadas no rebasan los 19 mg/m³.

Para la población en general, el humo de cigarrillo y los alimentos ahumados constituyen las fuentes más importantes de exposición al fenol, aparte de la exposición a través del aire. La exposición a través del agua potable y de los alimentos contaminados por inadvertencia probablemente sea baja; el fenol tiene un olor y un sabor desagradables, que en la mayor parte de los casos provocan el rechazo del consumidor.

5. Cinética y metabolismo

El fenol se absorbe fácilmente por todas las vías de exposición. Tras la absorción, la sustancia se distribuye rápidamente a todos los tejidos.

El fenol absorbido se conjuga principalmente con el ácido glucurónico y el ácido sulfúrico y, en menor medida, se hidroxila en pirocatequina e hidroquinona. También se conjuga con los fosfatos. La formación de metabolitos reactivos (4,4-bifenol y difenoquinona) se ha demostrado en estudios *in vitro* con neutrófilos y leucocitos humanos activados.

Las cantidades relativas de glucurono y sulfoconjugados varían según la dosis y la especie animal. Tras aumentar la dosis de fenol, se observó en las ratas un cambio de la sulfatación a la glucuronidación.

El hígado, los pulmones y la mucosa gastrointestinal constituyen los sitios más importantes del metabolismo fenólico. La función relativa desempeñada por esos tejidos depende de la vía de administración y de la dosis.

Estudios *in vivo* e *in vitro* han demostrado la unión covalente del fenol con las proteínas tisulares y plasmáticas. Algunos metabolitos fenólicos se unen asimismo a las proteínas.

La excreción por la orina es la principal vía de eliminación del fenol en los animales y en los seres humanos. La tasa de excreción urinaria varía en función de la dosis, de la vía de administración y de la especie. Una parte menor se excreta a través de las heces y del aire espirado.

6. Efectos en mamíferos de laboratorio y en sistemas de prueba *in vitro*

El fenol tiene una toxicidad aguda moderada en los mamíferos. En los roedores, los valores de la DL_{50} por vía oral oscilan entre 300 y 600 mg de fenol/kg de peso corporal. Los valores de la DL_{50} por vía cutánea para ratas y conejos oscilan entre 670 y 1400 mg/kg de peso corporal, respectivamente, y el valor de la CL_{50} por inhalación a las 8 horas en las ratas es superior a los 900 mg de fenol/m^3. Los síntomas clínicos después de la exposición aguda son hiperexcitabilidad neuromuscular y convulsiones graves, necrosis de la piel y de las mucosas de la garganta y efectos en los pulmones, fibras nerviosas, riñones, hígado y en la respuesta pupilar a la luz.

Las soluciones de fenol son corrosivas para la piel y los ojos. Los vapores de fenol pueden irritar las vías respiratorias. Existen pruebas de que el fenol no produce sensibilización cutánea.

Los efectos más importantes notificados a partir de estudios de corta duración en animales fueron neurotoxicidad, lesiones hepáticas y renales, trastornos respiratorios y retraso del crecimiento. Se han notificado efectos tóxicos en el riñón de las ratas con dosis por vía oral de 40 mg/kg al día o más. La toxicidad en el hígado resultó evidente en las ratas a las que se habían administrado al menos 100 mg/kg diarios. En un estudio limitado de 14 días de duración realizado en ratas se notificó un nivel sin efectos adversos observados (NOAEL) de 12 mg/kg al día por vía oral, basado en los efectos renales. En este experimento, la miosis (respuesta del iris a la luz) se mantuvo inhibida con 4 mg/kg al día; sin embargo, no está claro el significado médico de este hallazgo. Se notificó la existencia de algunos cambios biológicos en la mucosa intestinal y los riñones de ratones con dosis inferiores a 1 mg/kg al día, dato de significado toxicológico incierto.

No hay estudios adecuados sobre la toxicidad reproductiva del fenol. En estudios con ratas y ratones el fenol ha sido identificado como tóxico del desarrollo. En dos estudios de dosis múltiples en ratas, se han notificado NOAEL de 40 mg/kg al día (el más bajo nivel sin efectos adversos observados (LOAEL) fue de 53 mg/kg al día) y de 60 mg/kg al día (el LOAEL fue de 120 mg/kg al día). En el ratón, el NOAEL fue de 140 mg/kg al día (el LOAEL fue de 280 mg/kg al día).

La mayor parte de las pruebas de mutagenicidad bacteriana han dado resultados negativos. En células *in vitro* de mamíferos se han observado mutaciones, lesiones cromosómicas y efectos en el ADN. El fenol no tiene efectos en la comunicación intercelular (medida como cooperación metabólica) en cultivos de células de mamíferos. En algunos estudios se ha observado la inducción de micronúcleos en células de médula ósea de ratones. Con dosis más bajas no se observaron micronúcleos en estudios con ratones.

Se han llevado a cabo dos estudios de carcinogenicidad con ratas y ratones machos y hembras a los que se administró fenol con el agua de beber. Sólo se observó malignidad (por ejemplo, carcinoma de células C de la tiroides y leucemia) en ratas macho con dosis bajas. No se han realizado estudios adecuados de carcinogenicidad por vía dérmica o por inhalación. Estudios de carcinogenicidad de dos fases han mostrado que el fenol, aplicado repetidamente a la piel del ratón, tiene efectos activadores.

7. Efectos en el ser humano

Se ha notificado una larga serie de efectos adversos en el ser humano resultantes de la exposición bien documentada al fenol por vía cutánea, oral o intravenosa. Se ha notificado irritación gastrointestinal tras su ingestión. Los efectos locales de la exposición cutánea van desde el emblanquecimiento o el eritema indoloros hasta la corrosión y la necrosis profunda. Entre los efectos sistémicos cabe citar disritmias, acidosis metabólica, hiperventilación, disnea, insuficiencia renal aguda, lesiones renales, orinas oscuras, metahemoglobinemia, trastornos neurológicos (incluidas convulsiones), choque cardiovascular, coma y muerte. La dosis mínima reportada como causante de muerte en el ser humano es de 4,8 g por ingestión; la muerte se produjo en menos de 10 minutos.

Durante mucho tiempo se ha reconocido la posibilidad de envenenamiento por inhalación de los vapores de fenol, pero no se han reportado casos mortales relacionados con esta vía de exposición. Los síntomas que se asocian a la inhalación de fenol consisten, entre otros, en anorexia, pérdida de peso, dolor de cabeza, vértigo, salivación y orinas oscuras.

El fenol no es un agente sensibilizante.

El umbral de percepción del fenol por el olfato humano oscila entre 0,021 y 20 mg/m^3 en el aire. Se ha notificado un umbral de percepción del fenol en el agua de 7,9 mg/litro, y un umbral de percepción por el gusto de 0,3 mg/litro en el agua.

No se dispone de datos adecuados sobre la carcinogenicidad del fenol en el ser humano.

8. Efectos en los seres vivos del medio ambiente

En estudios con bacterias de especie única, los valores de la CE_{50} con inhibición del crecimiento oscilaron entre 244 y 1600 mg de fenol/litro. Se comprobó un umbral de toxicidad de 64 mg de fenol/litro. Los valores para los protozoarios y los hongos fueron de la misma cuantía que para las bacterias, mientras que para las algas fueron ligeramente inferiores.

El fenol es tóxico para los organismos superiores de agua dulce. Los valores más bajos de la CL_{50} o la CE_{50}, para crustáceos y peces se sitúan entre 3 y 7 mg de fenol/litro. Los datos sobre la toxicidad aguda para organismos marinos son comparables a los correspondientes a organismos de agua dulce. En estudios de larga duración sobre especies de crustáceos y de peces se han observado notables diferencias de sensibilidad; los valores de la CL_1 en pruebas con embriones y larvas de *Salmo* y *Carassius* resultaron mucho más bajos (0,2 y 2 μg de fenol/litro, respectivamente) que los valores correspondientes a otras especies de peces (NOLC de 2,2-6,1 mg/litro) y anfibios, o que los obtenidos en pruebas de reproducción en crustáceos (NOLC de 10 mg de fenol/litro). No se dispone de datos acerca de pruebas de larga duración realizadas en organismos marinos.

Los factores de bioconcentración del fenol en diversos tipos de organismos acuáticos son en general muy bajos (< 1-10), aunque se han notificado también algunos valores más altos (hasta 2200). Así pues, no se prevé que la bioacumulación del fenol sea significativa.

Los datos sobre el destino y los efectos del fenol en organismos terrestres son muy escasos. En el mijo se determinó una CE_{50} a las 120 horas de 120-170 mg/litro mientras que, en una prueba de contacto, la CL_{50} para especies de lombrices resultó ser de 2,4-10,6 μg/cm^2.

9. Resumen de la evaluación

9.1 Salud humana

La población en general está expuesta al fenol principalmente por inhalación. La exposición repetida por vía oral puede producirse por el consumo de alimentos ahumados o de agua potable.

No existen datos suficientes para determinar el grado de exposición de la población en general, pero se puede calcular la cantidad máxima ingerida diariamente. Basándose en «la peor de las hipótesis» se puede realizar una estimación suponiendo que un individuo estará expuesto en grado máximo al fenol mediante la inhalación continua de aire intensamente contaminado acompañada de un consumo frecuente de productos alimenticios ahumados y de agua que contenga fenol hasta niveles de percepción por el gusto. En total, la ingesta máxima diaria de fenol en un individuo de 70 kg se calcula en 0,1 mg/kg de peso corporal al día.

Los valores de NOAEL más bajos identificados en experimentos con animales se refieren a efectos en el riñón y en el desarrollo, y en las ratas se sitúan dentro de un margen de variación de 12-40 mg/kg de peso corporal al día. Utilizando un factor de incertidumbre de 200, se recomienda como límite máximo de la ingesta diaria total (IDT) entre 60 y 200 μg/kg de peso corporal al día. Teniendo en cuenta que el límite máximo de la ingesta diaria en seres humanos se calcula en 100 μg/kg de peso corporal al día, se llega a la conclusión de que la exposición media de la población en general al fenol, sea cual fuere la fuente, se encuentra por debajo de este intervalo.

Son motivo de preocupación algunas indicaciones de que el fenol podría ser genotóxico y el hecho de que no haya datos suficientes para descartar con seguridad la posible carcinogenicidad del compuesto. La evaluación debe mantenerse sujeta a revisión periódica.

9.2 Medio ambiente

No se prevé una bioacumulación importante del fenol. Este compuesto es tóxico para los organismos acuáticos; mediante la aplicación del método modificado de la Agencia de los EE.UU. para la Protección del Medio Ambiente, se puede determinar un nivel en medio ambiente de preocupación de 0,02 μg/litro. Se carece de datos adecuados sobre plantas y organismos terrestres.

El transporte de fenol entre compartimientos se produce principalmente por deposición hídrica y filtración a través del suelo. En general, es poco probable que el compuesto persista en el medio ambiente. La escasez de datos sobre la exposición no permiten evaluar el riesgo que representa el fenol para los ecosistemas tanto acuáticos como terrestres. Sin embargo, habida cuenta del nivel de preocupación ambiental que se ha establecido en relación con el agua, es razonable suponer que los organismos acuáticos pueden correr riesgo en cualquier agua superficial o marina contaminada con fenol.